T0205697

RESEARCH METHODOLOGY IN PHYSICS AND CHEMISTRY OF SURFACES AND INTERFACES

RESEARCH METHODOLOGY IN PHYSICS AND CHEMISTRY OF SURFACES AND INTERFACES

Edited by
Nekane Guarrotxena, PhD

A. Pourhashemi, PhD, Gennady E. Zaikov, DSc, and A. K. Haghi, PhD
Reviewers and Advisory Board Members

Apple Academic Press

TORONTO   NEW JERSEY

Apple Academic Press Inc.   Apple Academic Press Inc.
3333 Mistwell Crescent   9 Spinnaker Way
Oakville, ON L6L 0A2   Waretown, NJ 08758
Canada   USA

©2015 by Apple Academic Press, Inc.

First issued in paperback 2021

Exclusive worldwide distribution by CRC Press, a member of Taylor & Francis Group
No claim to original U.S. Government works

ISBN 13: 978-1-77463-332-8 (pbk)
ISBN 13: 978-1-77188-011-4 (hbk)

Library of Congress Control Number: 2014941340

Library and Archives Canada Cataloguing in Publication

Research methodology in physics and chemistry of surfaces and interfaces/edited by Nekane Guarrotxena, PhD; A. Pourhashemi, PhD, Gennady E. Zaikov, DSc, and A.K. Haghi, PhD,
Reviewers and Advisory Board Members.

Includes bibliographical references and index.
ISBN 978-1-77188-011-4 (bound)
1. Surface chemistry. 2. Nanostructured materials--Surfaces. 3. Chemistry, Physical and theoretical. I. Guarrotxena, Nekane, editor

QD506.R48 2014   541'.33   C2014-903627-2

Apple Academic Press also publishes its books in a variety of electronic formats. Some content that appears in print may not be available in electronic format. For information about Apple Academic Press products, visit our website at **www.appleacademicpress.com** and the CRC Press website at **www.crcpress.com**

ABOUT THE EDITOR

Nekane Guarrotxena, PhD

Nekane Guarrotxena, PhD, is Head of the Division in the Institute of Polymer Science and Technology in Madrid, Spanish National Research Council, Spain. She is a well-known scientist in the field of organic chemistry, chemistry and physics of polymers, and composites and nanocomposites. She has published 10 books and volumes and about 500 original papers and reviews. She was the Vice-Director of the Institute of Polymer Sciences and Technology from 2001 through 2005. From 2008 through 2011, she was a visiting professor in the Department of Chemistry, Biochemistry and Materials at the University of California, Santa Barbara, and the Center for Chemistry at the Space-Time Limit (CaSTL) at the University of California, Irvine. She is an editorial board member of the *Polymer Research Journal* and *ISRN* journals. Her research interests focus on the synthesis and assembly of hybrid nanomaterials, nanoplasmonics, and their uses in nanobiotechnology applications, such as bioimaging, drug delivery, therapy, and biosensing.

REVIEWERS AND ADVISORY BOARD MEMBERS

A. Pourhashemi, PhD

Dr. Ali Pourhashemi is Professor of chemical and biochemical engineering at Christian Brothers University in Memphis, Tennessee. He was formerly an Associate Professor and Department Chair there and also taught at Howard University in Washington, DC. He is a member of several professional organizations on the international editorial review board of the *International Journal of Chemoinformatics and Chemical Engineering*, and has published many articles and presented at many professional conferences.

Gennady E. Zaikov, DSc

Gennady E. Zaikov, DSc, is Head of the Polymer Division at the N. M. Emanuel Institute of Biochemical Physics, Russian Academy of Sciences, Moscow, Russia, and Professor at Moscow State Academy of Fine Chemical Technology, Russia, as well as Professor at Kazan National Research Technological University, Kazan, Russia. He is also a prolific author, researcher, and lecturer. He has received several awards for his work, including the Russian Federation Scholarship for Outstanding Scientists. He has been a member of many professional organizations and on the editorial boards of many international science journals.

A. K. Haghi, PhD

A. K. Haghi, PhD, holds a BSc in urban and environmental engineering from University of North Carolina (USA); a MSc in mechanical engineering from North Carolina A&T State University (USA); an DEA in applied mechanics, acoustics and materials from Université de Technologie de Compiègne (France); and a PhD in engineering sciences from Université de Franche-Comté (France). He is the author and editor of 65 books and 1000 published papers in various journals and conference proceedings. Dr. Haghi has received several grants, consulted for a number of major corporations, and is a frequent speaker to national and international audiences. Since 1983, he served as a professor at several universities. He is currently Editor-in-Chief of the *International Journal of Chemoinformatics and Chemical Engineering* and *Polymers Research Journal* and on the editorial boards of many international journals. He is a member of the Canadian Research and Development Center of Sciences and Cultures (CRDCSC), Montreal, Quebec, Canada.

CONTENTS

LIST OF CONTRIBUTORS

A. A. Bokarev
Moscow State University of Applied Biotechnology, Moscow, Russia

W. W. Focke
Institute of Applied Materials, Department of Chemical Engineering, University of Pretoria, Pretoria 0002, South Africa
Tel: (+27) 12 420 3728, Fax: (+27) 12 420 2516

D. Gnanasekaran
Institute of Applied Materials, Department of Chemical Engineering, University of Pretoria, Pretoria 0002, South Africa
Tel: (+27) 12 420 3728, Fax: (+27) 12 420 2516

N. Guarrotxena
Instituto de Ciencia y Tecnología de Polímeros (ICTP) , Consejo Superior de Investigaciones Científicas (CSIC) , Juan de la Cierva 3, 28006 Madrid, Spain
E-mail: nekane@ictp.csic.es

A. K. Haghi
University of Guilan, Rasht, Iran

H. Hlídková
Institute of Macromolecular Chemistry Academy of Sciences of the Czech Republic, v.v.i.
Heyrovský Sq. 2, 162 06 Prague 6, Czech Republic

D. Horák
Institute of Macromolecular Chemistry Academy of Sciences of the Czech Republic, v.v.i.
Heyrovský Sq. 2, 162 06 Prague 6, Czech Republic
E-mail address: horak@imc.cas.cz

V. A. Ilatovsky
N.N. Semenov Institute of Chemical Physics, Russian Academy of Sciences, 4 Kosygin Street, Moscow 119991, Russia

S. G. Karpova
Emanuel Institute of Biochemical Physics, Russian Academy of Sciences, Moscow, Russia

G. G. Komissarov
N.N. Semenov Institute of Chemical Physics, Russian Academy of Sciences, 4 Kosygin Street, Moscow 119991, Russia
E-mail: gkomiss@yandex.ru; komiss@chph.ras.ru

G. A. Korablev
Izhevsk State Agricultural Academy, Russia, Izhevsk 426000
E-mail: korablev@udm.net

N. G. Korableva
Izhevsk State Agricultural Academy, Russia, Izhevsk 426000
E-mail: korablev@udm.net

O. A. Legonkova
Moscow State University of Applied Biotechnology, Moscow, Russia
E-mail: OALegonkovaPB@mail.ru

P. H. Massinga Jr
Universidade Eduardo Mondlane, Faculdade de Ciências, Campus Universitário Principal, Av. Julius Nyerere, P.O. Box 257, Maputo, Moçambique

M. Mudarra
Dept. Física i Enginyeria Nuclear. ETSEIAT, Universitat Politècnica de Catalunya,Colom, 11 Terrassa 08222, Barcelona, Spain

A. A. Popov
Emanuel Institute of Biochemical Physics, Russian Academy of Sciences, Moscow, Russia

A. Pourhashemi
Department of Chemical and Biochemical Engineering, Christian Brothers University, Memphis, Tennessee, USA

G. A. Ptitsyn
N.N. Semenov Institute of Chemical Physics,Russian Academy of Sciences, 4 Kosygin Street, Moscow 119991, Russia

G. V. Sinko
N.N. Semenov Institute of Chemical Physics, Russian Academy of Sciences, 4 Kosygin Street, Moscow 119991, Russia

G. E. Zaikov
Institute of Biochemical Physics N. M. Emanuel, Russian Academy Sciences, 4 Kosygina Street, Russia, Moscow 119991
E-mail: chembio@sky.chph.ras.ru

LIST OF ABBREVIATIONS

1DNSMs	one-dimensional nanostructures
2DNSMs	2-D nanostructures
CXR	cyclohexane regain
DSC	differential scanning calorimeter
EHD	electrohydrodynamic
EPR	electron paramagnetic resonance
eV	electron volts
EVA	(ethylene-*co*-vinyl acetate
FNI	Mozambican Research Foundation
HOMO	highest occupied state
HRR	heat release rate
HVSEM	high-vacuum SEM
IRDP	Institutional Research Development Programme
LDHs	layered double hydroxides
LUMO	lowest free state
LVSEM	low-vacuum scanning electron microscopy
MFI	melt flow index
NRF	National Research Foundation
NSMs	nanostructured material
Pcs	phthalocyanines
PMMA	poly(methyl methacrylate)
POSS	polyhedral oligomericsilsesquioxane
SEC	size exclusion chromatography
TEG	thermal expanded graphite
TEM	transmission electron microscopy
THF	tetrahydrofuran
THR	total heat release
TPCs	tetrapyrrole compounds
TPP	tetraphenylporphyrin
VA	vinyl acetate
WL	weight loss
XRD	X-ray diffraction analysis

LIST OF SYMBOLS

h	Planck constant
I_{ph}	photocurrent (μA)
W	absorbed light power (W)
λ	wavelength (m)
c	speed of light (m/s)
ϑ	total concentration of adsorbed molecules
W	activation energy of the transition to the charged form
E_a	distance from the surface acceptor level to the E_v
F_s	distance from the Fermi level at the surface to E_v
E_v	ceiling of the valence band
w_w	weight of hydrated sample
w_d	weight of dry sample
p	porosity of the hydrogels
W	bond energy of electrons
E_i	ionization energy of an atom
W_i	bond energy of an electron
n_i	number of elements of the given orbital
r_i	orbital radius of i orbital
Z^* and n^*	effective charge of a nucleus and effective main quantum number
P_0	spatial-energy parameter
P_E	value of effective P-parameter
P_C	parameter of a complex structure
N_1 and N_2	number of homogeneous atoms
N	bond order
K	maxing or hybridization coefficient
D_0	dissociation energy of a molecule
K_E	equilibrium constants of chemical reaction
T	thermodynamic temperature of the process
$\Delta \Phi_T^*$	change in the reaction of considered reduced thermodynamic potential
R	universal gas constant
ΔH_G^0	formation enthalpy of gaseous substance

ΔH^0_{SD}	formation enthalpy of solid in nonstandard state
ΔH_S	sublimation enthalpy
R	radius
n	main quantum number
r	dimensional characteristic of atom structure
N_0	number of particles in the sphere volume of the radius R
σ_0	DC conductivity
A	temperature-dependent parameter
n	fractional exponent
t^*	relaxation time
T_g	glass transition temperature
$\varepsilon\varepsilon_0$	ratio of the dielectric permeability
ρ^e	local free charge density
E	electric field strength
P	polarization
Q	dipole charge
d	orientation
δ	two-dimensional vector
σ	the surface charge density; the stress on the element
R_1 and R_2	principal radii of curvature
α	surface tension coefficient
a	average radius of a_{an} segment of the jet between two beads
h	distance
ε	strain of the element
μ	viscosity of the material

PREFACE

The aim of this book is to provide both a rigorous view and a more practical, understandable view of the interfaces of physics and chemistry in micro and nanoscale materials. This book intends to satisfy readers who have both direct and lateral interest in the discipline.

This volume is structured into different parts devoted to micro and nanostructured systems and their applications. Every section of the book has been expanded where relevant to take account of significant new discoveries and realizations of the importance of key concepts. Further, emphases are placed on the underlying fundamentals and on acquisition of a broad and comprehensive grasp of the field as a whole.

This book brings together research contributions from eminent experts on subjects that have gained prominence in material and chemical engineering and science. The uniqueness of the topics presented in the book can be gauged from their fundamental and practical importance, particularly the latest developments in advanced materials chemical domains. With contributions from experts from both industry and academia, this book presents the latest developments in the identified areas. This book incorporates appropriate case studies, explanatory notes, and schematics for more clarity and better understanding.

This book will be useful for chemists, chemical engineers, technologists, and students interested in this research and its applications.

This new book

- features classical topics that are conventionally considered as part of physical chemistry.
- covers the physical chemistry principles from a modern viewpoint.
- focuses on topics with more advanced methods.
- emphasizes on precise mathematical development and actual experimental details.

— **Nekane Guarrotxena, PhD**

CHAPTER 1

HIGHLY FILLED COMPOSITE MATERIALS WITH REGULATED PHYSICAL AND MECHANICAL PROPERTIES BASED ON SYNTHETIC POLYMERS AND ORGANIC AND INORGANIC FILLERS

O. A. LEGONKOVA, A. A. POPOV, A. A. BOKAREV, and S. G. KARPOVA

CONTENTS

1.1 AIMS AND BACKGROUNDS

The aim of this study was to upgrade the highly filled composite materials with regulated physicomechanical properties based on synthetic polymers and organic and inorganic fillers that can be used in the creation of environment-friendly biodegradable products for different purposes. Features that are essential for technology, such as physical, chemical, structural, and rheological properties, were investigated. Electron paramagnetic probe and IR-spectrometry methods were used to prove that inorganic filler plays the role of plasticizer in the creation of hybrid composites. Microbiological aspects will be discussed in another article.

1.2 INTRODUCTION

One of the promising trends from the viewpoint of ecology is the development of biodegradable polymer composites. These materials, along with the polymer base more resistant to biodegradation, comprise fillers that are not only accessible for microbial degraders but are also agroindustrial wastes to be utilized. [1] Search for cheap fillers and development of polymer composites makes it possible not only to reduce the cost of product, but it is also a solution to ecological problems.

1.3 EXPERIMENTAL

Third-grade threshed grain wastes (size of particles, 63–240 µm; bulk density, 350 kg/m³; and humidity, 4%) were used as organic filler. As inorganic filler, a Rastvorin-A water-soluble mineral fertilizer (OST 10-193-96, produced by the Buysk Mineral Fertilizer Plant) was taken in the following composition (in %): $(NH_4)_2SO_4$, 35; $NH_4H_2PO_4$, 6; KNO_3, 32; and $MgSO_4 \cdot 7H_2O$, 27.

The physicomechanical properties of specimens within a broad range of ingredient ratios were determined according to GOST 14236-82; the rheological characteristics of filled compositions were determined by the capillary viscosimetry method.

In this study, electron paramagnetic probe method was used, which was a stable nitroxyl radical 2,2,6,6-tetramethylpiperidine-1-oxyl. The radical was introduced into films from vapors at $T = 25°C$ up to a concentration of 10^{-3}

mol/L. A reference solution of the radical in CCl_4 with a known number of spins was used for determining the concentration of radicals in films. The number of spins in a specimen was determined by comparing the areas under the absorption curves of the specimen studied and the reference. The rotational mobility of the probe was characterized by the correlation time τ. The values of τ were assessed from the electron paramagnetic resonance (EPR) spectra by the method outlined by Antsyferova et al. [2]

1.4 RESULTS AND DISCUSSIONS

While introducing fillers into SEVA irrespective of its grade, it was found that the strength and relative elongation decreased with an increase in filler content in both organic and inorganic fillers in two-component polymer–filler systems (see Figures 1 and 2). Introduction of organic filler made the specimens more rigid; with the introduction of inorganic filler even at a high concentration (60 wt %), specimens preserve a high plasticity (the breaking strain is 400%).

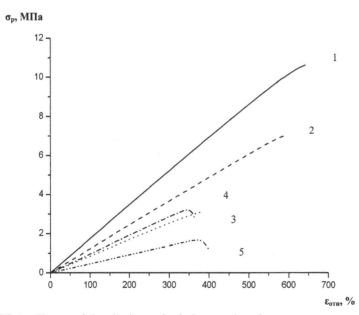

FIGURE 1 Change of the physicomechanical properties of a two-component system (SEVA/inorganic filler, wt %): 1, 100/0; 2, 80/20; 3, 60/40; 4, 50/50; and 5, 40/60.

In the case of three-component systems too (SEVA/inorganic filler/organic filler, Figure 3), an increase in the content of inorganic filler leads to more plastic specimens.

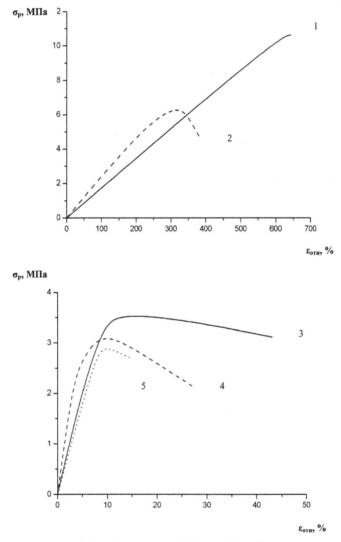

FIGURE 2 Change of the physicomechanical properties of a two-component system (SEVA/organic filler, wt %): 1, 100/0; 2, 80/20; 3, 60/40; 4, 50/50; and 5, 40/60.

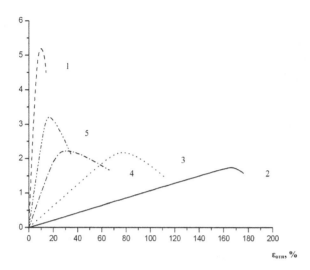

FIGURE 3   The physicomechanical properties of a three-component system based on SEVA-113, wt %: 1, SEVA/organic filler, 50/50; 2, SEVA/inorganic filler, 50/50; 3, SEVA/inorganic filler/organic filler, 50/37/13; 4, SEVA/inorganic filler/organic filler, 50/25/25; and 5, SEVA/inorganic filler/organic filler, 50/13/37.

Figure 4 presents electron micrographs of the fracture faces of composites. The systems are considered to be heterogeneous. The systems with inorganic filler include both crystallites of filler salts and their more complex formations. Defects in the system are "pressed into" the homogeneous structure of the composite. For this reason, a decrease of strength of polymer–composite material occurs, owing to a decrease of the content of polymer in the composition.

The EPR method supports the data obtained (Table 1): the higher the concentration of organic wastes in a hybrid composite, the greater the correlation time, that is, the more rigid the system is, the smaller the deformability of the system. At the same time, the higher the concentration of inorganic filler, the more plastic the systems are.

Noteworthy is the behavior of two ternary systems: 50 percent sevilene at 25/25 of organic and inorganic filler and 15 percent sevilene at 50/35 of organic and inorganic filler, for which the correlation time of all ternary systems studied is minimal. Both the systems are plastic and capable of being processed by commercial equipment into particular articles.

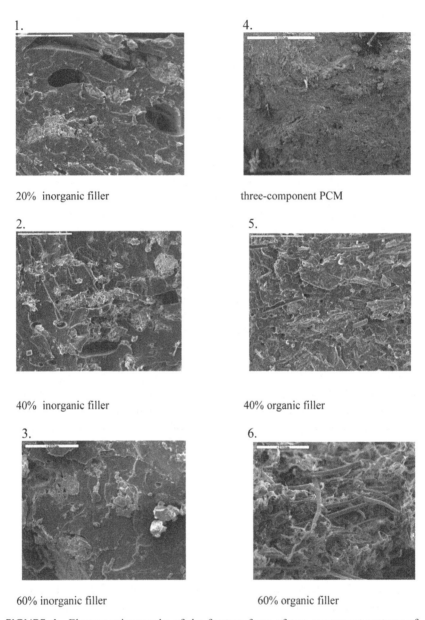

1.

20% inorganic filler

4.

three-component PCM

2.

40% inorganic filler

5.

40% organic filler

3.

60% inorganic filler

6.

60% organic filler

FIGURE 4 Electron micrographs of the fracture faces of two-component systems of polymer–composite material based on sevilene and inorganic filler (1, 2, 3); sevilene and organic filler (5, 6); three-component PCM (4): 15 percent sevilene, 50 percent organic filler, 35 percent inorganic filler (magnification ×200 μm).

TABLE 1   Correlation time of the paramagnetic probe in PCM specimens depending on the extent of filling with various fillers

Extent of filling with organic filler (%)	τ	Extent of filling with inorganic filler (%)	τ	Extent of filling with organic/inorganic filler (%)	τ
10	$6.1 \cdot 10^{-10}$	20	$7.7 \cdot 10^{-10}$	18.75/18.75	$6.1 \cdot 10^{-10}$
15	$6.8 \cdot 10^{-10}$	40	$7.4 \cdot 10^{-10}$	25/25	$5.1 \cdot 10^{-10}$
20	$7.1 \cdot 10^{-10}$	50	$7.1 \cdot 10^{-10}$	50/25	$6.2 \cdot 10^{-10}$
35	$6.3 \cdot 10^{-10}$	60	$6.8 \cdot 10^{-10}$	60/25	$6.5 \cdot 10^{-10}$
40	$7.0 \cdot 10^{-10}$	80	$6.4 \cdot 10^{-10}$	30/35	$6.0 \cdot 10^{-10}$
50	$7.1 \cdot 10^{-10}$	–	–	50/35	$6.2 \cdot 10^{-10}$
60	$8.0 \cdot 10^{-10}$	–	–	50/35	$5.7 \cdot 10^{-10}$

As inorganic filler is a mixture of salts (potassium, and magnesium chlorides; magnesium sulfates; and potassium and ammonium nitrates), the nitrogen-containing salts led to a significant increase of the melt flow index in sevilene-based PCMs (Table 2).

TABLE 2   Effect of the composition and concentration of inorganic salts on the melt flow index (g/10 min) of sevilene-based composites

Content (wt %)	Unfilled specimen	$MgCl_2$	KCl	$MgSO_4$	KNO_3	NH_4NO_3
1	7.8	7.5	6.7	8.1	2.9	4.3
2	7.5	7.1	6.3	7.1	Too high flow rate made its measurement impossible	

To assess the possibility of chemical interaction of fillers with polymers, we used the method of IR spectroscopy. The following IR bands were found to be characteristic of SEVA: 1740 cm^{-1} (vibrations of the aldehyde group), 1380 cm^{-1} (vibrations of the vinyl acetate group), 1240 cm^{-1}, 1050 cm^{-1}, 797 cm^{-1}, and 607 cm^{-1} [3].

According to Nakanishi [3], vibrations in the region of 1440 cm^{-1} can be assigned to

The methylene groups $-CH_2-$ in SEVA have characteristic valent vibrations at 2850 cm^{-1}. The NH_4^+ group has the absorption band in the region of 3,300–3,030 cm^{-1}; the group $P-O-C_{alkyl}$, 1,180–1,150 cm^{-1}; the groups SO_4^{2-}, 1,130–1,080 cm^{-1}; $S-CH_2$, 2,700–2,630 cm^{-1}; the groups PO_4^{3-}, HPO_4^{2-}, $H_2PO_4^-$, in the region of 1,100–1,000 cm^{-1}; and NO_3^-, 1,380 cm^{-1}.

Ratios of the optical densities of the bands 1,440 cm^{-1} and 1,380 cm^{-1} to the optical density of the band 2,850 cm^{-1} taken as the internal standard of the specimens of mixtures of SEVA with the salts KNO_3, NH_4NO_3, and K_2HPO_4 shown in Table 3.

The table shows that while heating the salts KNO_3, NH_4NO_3, and K_2HPO_4, there is an insignificant splitting-off of the vinyl acetate group from the main chain. Heating of polymer with K_2HPO_4 produces a band responsible for the interaction of $P-O-C_{alkyl}$ ($D_{1180} = 0.62$).

In the case of the other salts (KH_2PO_4, $(NH_4)_2HPO_4$, $(NH_4)_2SO_4$, $MgSO_4$), such an effect could not be identified due to the coincidence of the spectra of the characteristic absorption bands.

TABLE 3 Ratios of the optical densities of the bands 1,440 cm^{-1} and 1,380 cm^{-1} to the optical density of the band 2,850 cm^{-1} taken as the internal standard of the specimens of mixtures of SEVA with the salts KNO_3, NH_4NO_3, and K_2HPO_4

	Sevilene	Sevilene + KNO$_3$	Sevilene + NH$_4$NO$_3$	Sevilene + K$_2$HPO$_4$	Sevilene + inorganic filler
D_{1440}/D_{2850}	0.66	0.47	0.46	0.42	0.41
D_{1380}/D_{2850}	0.59	0.42	0.43	0.31	0.36

Table 4 presents the melting heats of polymers and polymer–composite materials at temperatures of solid-phase transitions of nitric salts, coinciding with sevilene fluidity temperature of 80°C. According to Chemist's Reference Book, Moscow [4], ammonium nitrates have solid-phase transitions at temperatures of 80°C and 130°C with ΔH_m equal to 0.32 ccal/mol and 1.01 ccal/mol, respectively; the salt KNO_3 has the solid-phase transition at 130°C with ΔH_m equal to 1.3 ccal/mol.

TABLE 4 Melting heats of polymers and their composites at temperatures of solid-phase transitions of nitric salts (J/g)

Material	80°C	130°C
Sevilene unfilled ($T_t = 80°C$)	49.9	–
KNO_3	–	54.2
NH_4NO_3	10.1	53.9
Sevilene filled with 20% NH_4NO_3	49.9	9.8
Sevilene filled with 20% KNO_3	44.9	7.6
Sevilene filled with 2% KNO_3	65.1	1.1
Sevilene filled with 40% inorganic filler	50.4	–
Sevilene/organic filler/inorganic filler (15:50:35)	–	94.9

At 80°C, for polymer–composite material based on sevilene an increase of melting heat at the introduction of 2 percent KNO_3 is due to the fact that densely packed structures are formed in polymer's liquid phase at a small content of solid filler at the interface with solid particles of filler KNO_3. [5] At an increase of the content of salt up to 20 percent, the decrease of the melting heat is due to a decrease of the content of polymer per 1 g of specimen studied.

At the introduction of NH_4NO_3 into this system, there is no change of melting enthalpy compared with pure polymer. At 130°C, when the temperature of the solid-phase transition of the salts is higher than the melting temperature of sevilene, the enthalpy of the composite decreases, which is due to a decrease of the content of salt per 1 g of specimen studied.

1.5 CONCLUSIONS

Thus, the thermal effect of two-component hybrid systems formed from the melt is minimal when T_m (T_t) of the polymer is equal to or lower than the solid-phase transition temperature of the salts. By the extent of a decrease of enthalpy of the measured component, an indirect judgment can be made on the looseness of the filled system when compared with the individual components of the composite.

The process properties of hybrid composites are related not only to the decrease of the concentration of the matrix in the bulk of the composite but also to the plasticizing action of inorganic filler.

KEYWORDS

- **Correlation time**
- **Hybrid composites**
- **Hybrid composites**
- **Industrial wastes**
- **Inorganic fillers**
- **Melt Flow index**
- **Organic fillers**
- **Physical and chemical properties**
- **Polymers**
- **Structure**

REFERENCES

1. Legonkova, O. A.; Biodeterioration of polymer-based composite materials in the environment. *Mater. Sci.* **2008**, *2*, 50–55.
2. Antsyferova, L. I.; Wasserman, A. M.; Ivanova, A. I.; Lifshits, V. A.; and Nazemets, N. S.; Atlas of the Spectra, Moscow: Nauka Publishers; **2012**, 160 (in Russian).
3. Nakanishi, K.; Infrared Spectra and Structure of Organic Compounds. Moscow: Mir Publishers; **2012**, 200 p.
4. Chemistry Handbook, Moscow–Leningrad: Khimiya Publishers; **2011**, *1*, 1070 p. (in Russian).
5. Bryk, M. T.; Utilization of Filled Polymers, Moscow: Khimiya Publishers; **2012**, 190 p. (in Russian).

CHAPTER 2

PREDICTION OF PHOTOELECTROCHEMICAL PROPERTIES OF SELECTED MOLECULES BY THEIR STRUCTURE

V. A. ILATOVSKY, G. V. SINKO, G. A. PTITSYN,
and G. G. KOMISSAROV

CONTENTS

2.1 AIM AND BACKGROUND

The aim of our study was to provide objective criteria to predict theoretically, at least qualitatively, photoelectrochemical properties of the molecules by their structure. In our opinion, the most important factor in the selection process is the electron density distribution in molecules.

2.2 INTRODUCTION

With the development of the photovoltaic and photoelectrochemical solar energy converters, interest in organic semiconductors is becoming an increasingly applied nature inherent in the transition to industrial development. At this stage, it is very important to select the most promising classes of organic semiconductors and even more important to provide objective criteria to predict theoretically, at least qualitatively, photoelectrochemical properties of the molecules through their structures. In our opinion, the most important factor in the selection process is the electron density distribution in molecules, as it determines both the individual properties of the specific compound (organic semiconductor, pigment, and dye) and the intermolecular interaction, that is, the behavior of the molecules in a solid after condensation. Good examples of powerful rearrangement of electronic structure and corresponding changes of physical and chemical properties of molecules can be observed in the well-known tetrapyrrole compounds (TPCs), which, of course, belong to the group of the most promising compounds. This is largely due to the enormous variability of the structure of TPC, typified by porphyrins and their derivatives, which are cyclic aromatic polyamines, conjugated with the multiloop system consisting of a 16-membered macrocycle with a closed π-conjugation system including 4–8 nitrogen atoms. By replacing hydrogen atoms in the pyrrole rings and mesopositions of various donor–acceptor groups, over 1,000 porphyrins have been produced. To this is added the variations due to the formation of different metal complexes and the introduction of extraligands. Substitution of CH bridges by nitrogen atom (aza-substitution) in TBP gives a representative group of phthalocyanines (Pcs) (else, tetrabenzo-porphyrazines/tetra-(butadiene-1,3-ylene-1,4) tetraazaporphin) and their derivatives.

Choosing the most effective working pigment from the various groups is extremely difficult, especially as the theoretical assumptions for this choice are almost none. At certain stages, one has to use intuitively guided screening to detect certain patterns that will eventually lead to the creation of a more or less acceptable theory. Thus, in comparison with the photocatalytic activity (the photovoltaic Becquerel effect) of thin films of TPC with different macrocycle structure, it was noted that the maximum photocurrent (I_{ph}) and photopotential (U_{ph}) are given by pigments, whose macrocycle's electronic structure modification is caused by exposure to a carbon atom in meso-position.[1] For example, tetraphenyl porphyrin (TPP) substitution of $\rangle$á−, â−, ã−, ä −$\langle$ hydrogen to phenyl groups leads to a high transfer of electron density to the π-conjugation circuit (and hence on the pyrrole rings) and significantly increases the photoactivity of TPP. Additional polarization arises due to the direct dipole interaction of atoms in the β-positions with the phenyl group. Slightly higher photocatalytic activity was shown by phthalocyanines [2], but, unlike TPP, Pc nitrogen directly substitutes carbon atoms in the mesoposition. Further, Pc has more substituents—benzene rings conjugated with the pyrrole rings—hence, aromaticity of this compound is much higher, and it is difficult to determine which substitution (aza- or benzo-) causes an increase in photoactivity.

As the nitrogen atoms are likely to be key points of the TPC macrocycle structure, there was an intention to conduct a sequential aza-substitution of all four CH groups in the TBP molecule to form phthalocyanine, with a fairly representative group of metal complexes with different degrees of the ligand bond ionicity, and explore changing of the photoelectrochemical characteristics of the films of produced pigments. Hoping to get a large enough material for reflection, in parallel with the experiments, quantum chemical calculations of the changes in the distribution of electron density, the bond order, and the energies of the orbitals in molecules of TPC were carried out. The calculations were performed by the program GAMESS, using Rutan molecular orbitals, by restricted Hartree–Fock method (not taking into account the correlation effects) and by density functional theory method (approximately takes into account the correlation effects). Both experimental and calculated data have shown a good correlation with phenomenological assumptions.

2.3 EXPERIMENTAL

In setting up experiments comparing the PEC properties of various aza-derivatives of TBP and their metal complexes, we focused on the adequacy of the conditions of measurement, reproducibility, and statistical significance of the results. For each modification of a pigment, there were measured parameters of 30 electrodes and already performed statistical analysis. The mean square variation of parameters in the cited experimental data was $\sigma = 5.6$ percent. Moreover, given the strong dependence of photocurrents I_{ph} (pH) and photo-potentials U_{ph} (pH) of pigmented electrodes on the pH of the electrolyte [3,4], extreme points were determined. Figure 1 shows the complete dependence of the potentials and currents of Pt-ZnPc electrode plotted from the average data for the 30 electrodes. For other pigments, similar dependencies were obtained, with which their photocatalytic activities were compared in the oxygen reduction reaction, wherein the first single-electron step is endothermic:

$$H^+ + O_2 + e^- \rightarrow HO_2, E_o = -0.32 \text{ V}$$

And this requires high activation energy. Excitons, photogenerated in the bulk of the film, migrate to the pigment–electrolyte interface. High concentrations of oxygen adsorbed on the film ($\sim 10^{15}$ cm^{-3}) provide necessary conditions for separation of electron–hole pairs arising from the collapse of the excitons due to efficient trapping of electrons with the formation of the charged form of adsorption O_2^-. The increase in the concentration of hydrogen ions promotes the completion of the reduction process and the transfer of the reaction products in the electrolyte. However, as can be noted from Figure 1, there is an optimal pH zone for I_{ph} (pH) and U_{ph} (pH), beyond which further increase in the concentration of H$^+$ leads to a decrease in the photoresponse. This is due to the different photoelectrochemical stability of pigments, particularly the possibility of protonation in the acidic environment, which is largely determined by the redox potential and ionization potential. In this case, the comparison was performed by photoactive maxima of I_{ph} (pH) and U_{ph} (pH).

Films of TBP and its metal derivatives with a thickness of 50 nm were applied to the polished platinum substrates with a diameter of 11 mm by thermal vacuum sublimation in quasi-closed volume [5–7], using a turbo-molecular vacuum system Varian Mini-TASK (vacuum up to 10^{-9} Torr) with the original design of evaporation chamber. Pigments used were previously purified twice by sublimation in vacuum upto 10^{-7} Torr. The amounts of impurities did not exceed 10^{-4} percent. The film thickness was controlled by frequency change of a quartz resonator located near the substrates in the evaporator system. Calibration curves were obtained by determining the thickness of pigmented films on transparent quartz substrates placed in the pigment's deposition zone, with a microinterferometer MII-11. Control of spectral parameters of the films confirmed integrity of the structure of pigment molecules. Because of the carefully refined method, we were able to obtain highly reproducible parameters of the films on platinum substrates. In particular, in the study of PEC parameters, variations in the photocurrents did not exceed 15 percent and the photopotentials up to 20 percent.

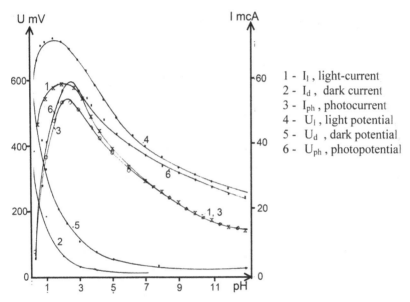

FIGURE 1   The dependence of currents and potentials of the electrode with Zn–Pc film on the Pt substrate on the pH of the electrolyte.

Photoactivity of the films were compared using Becquerel effect value; I_{ph} and U_{ph} were measured using a relatively saturated Ag/AgCl electrode in aerated 1.0 N KCl electrolyte, changing the pH from 1 to 14 by titration with 1.0 N KOH or HCl solution. Pigmented electrodes were illuminated with an arc xenon lamp DKSSh-120 with a stabilized power supply: the light output in the plane of the electrode was 100 mW/cm². For measurement of the quantum yield on a current η, pigment films with thickness d from 2 to 300 nm were applied on polished quartz substrates coated with a conductive layer of platinum. The η value was determined from a ratio

$$\eta = 6.25 \cdot 10^7 [hcI_{ph}/W\lambda] \ (\%)$$

where h—Planck constant, I_{ph}—photocurrent (μA), W—absorbed light power (W), λ— wavelength (m), and c—speed of light (m/s). Measurements were carried out in the region of maximum absorption of the pigments, separated from the spectrum of the lamp DKSSh-120 bandwidth of 10 nm using interference filters at luminous flux in the plane of the electrode 10 mW/cm^2.

2.4 RESULTS AND DISCUSSION

The results of measurement of photoelectrochemical activity of the films of 45 metal complexes of derivatives of tetraazabenzoporphyrin with varying degrees of azasubstitution are summarized in Table 1.

TABLE 1 Variations of photopotentials, photocurrents, and quantum yield on a current of thin film electrodes based on the aza-substituted metal complexes of tetrabenzo-porphyrin depending on the degree of substitution and the type of central atom

No.	TPC	U_{ph}	pH	I_{ph}	pH	η
1.	Mn–TBP	102	4	0.7	4	2.0
2.	Mn–MATBP	145	3	2.8	3	4.0
3.	Mn–DATBP	178	3	4.1	3	4.5
4.	Mn–TATBP	198	2	5.5	2	5.0
5.	Mn–Pc	230	1	6.5	1	6.0
6.	Ni–TBP	53	3	0.1	3	1.1
7.	Ni–MATBP	67	2	0.2	3	1.3
8.	Ni–DATBP	72	1	0.2	2	1.3
9.	Ni–TATB	78	1	0.3	1	1.4
10.	Ni–Pc	80	0	0.3	0	1.4
11.	Co–TBP	71	2	0.1	2	1.2
12.	Co–MATBP	92	1	0.2	1	1.5
13.	Cu–DATBP	105	1	0.3	1	1.5
14.	Co–TATBP	108	1	0.4	1	1.6

TABLE 1   *(Continued)*

No.	TPC	U_{ph}	pH	I_{ph}	pH	η
15.	Co–Pc	110	0	0.4	0	1.6
16.	Zn–TBP	308	5	5.2	5	9.0
17.	Zn–MATBP	431	4	20.0	4	13.0
18.	Zn–DATBP	500	3	32.0	3	15.0
19.	Zn–TATBP	540	2	46.0	2	17.0
20.	Zn–Pc	570	2	54.0	2	18.0
21.	Mg–TBP	160	4	4.0	3	4.0
22.	Mg–MATBP	280	4	9.0	3	6.0
23.	Mg–DATBP	320	3	14.0	2	7.0
24.	Mg–TATBP	370	2	17.0	1.5	9.0
25.	Mg–Pc	400	2	21.0	1.5	10.0
26.	H_2–TBP	162	4	1.0	3	2.0
27.	H_2–MATBP	227	3	4.0	2	4.0
28.	H_2–DATBP	260	2	6.2	2	5.0
29.	H_2–TATBP	285	1	7.7	1	6.0
30.	H_2–Pc	300	1	8.2	0	6.0
31.	Fe–TBP	85	3	0.4	2	1.0
32.	Fe–MATBP	140	1	1.1	1	3.0
33.	Fe–DATBP	155	1	1.2	0	3.2
34.	Fe–TATBP	170	0	1.3	0	3.5
35.	Fe–Pc	180	0	1.4	0	4.0
36.	ClFe–TBP	105	4	1.7	3	2.5
37.	ClFe–MATBP	180	3	3.5	2	3.8
38.	ClFe–DATBP	210	2	4.6	2	4.5
39.	ClFe–TATBP	240	1	5.0	1	5.0
40.	ClFe–Pc	250	1	5.4	1	5.0

TABLE 1   *(Continued)*

No.	TPC	U_{ph}	pH	I_{ph}	pH	η
41.	VO–TBP	240	4	11.0	3	5.0
42.	VO–MATBP	290	2	17.0	2	9.0
43.	VO–DATBP	330	1	21.0	1	13.0
44.	VO–TATBP	350	1	26.0	1	15.0
45.	VO–Pc	350	1	28.0	0	16.0

U_{ph}—maximum values of the photopotential in mV; I_{ph}—maximum photocurrent in μA; pH—value of pH at which the maximum of U_{ph},I_{ph}; η—quantum efficiency of a current; TBP—tetrabenzoporphyrin, MATBP—mono-tetraazabenzoporphyrin, DATBP—di-tetraazabenzoporphyrin, TATBP—tri-tetraazabenzoporphyrin and Pc—phthalocyanine (tetra-tetraazabenzoporphyrin).

For all studied compounds, the same effect of aza-substitution is observed; the most dramatic change of photocurrent (2–5 times) and photopotential (1.3–1.7 times) is the first step—the transition from the TBP to a mono-aza-substituted derivatives. Each subsequent step makes a smaller contribution, but the overall increase in the transition to the structure of phthalocyanine for U_{ph} reaches 185 percent, and for I_{ph} reaches 1000 percent. The stability of pigments changes significantly—the pH value, at which protonation of the molecules begins, decreases by 4–5 units as the proportion of nitrogen atoms in the macrocycle grows, which is associated with the increase of redox potential with aza-substitution. Changes in the stability of pigments are seen in the stability of the electrodes. Photoactivity of the TBP films is lost in 10–12 hours. At the same time, a transfer of dications of the pigments into the solution was spectrally recorded. Phthalocyanine films (regardless of the nature of the central atom) did not change their characteristics after hundreds of hours of light in similar conditions. Basically, the differences in the photoactivity of pigments are caused by three factors—the efficiency of energy (charge) transfer, the band gap, and the spectral range and extent of light absorption. All three factors are finally determined by the distribution of electron density in the molecules. Hence, for all of aza-derivatives of TBP, as a result of the transition from the structure of pure porphyrin to the phthalocyanine, an increase of photochromic properties of molecules and molecular interaction is observed. For example, Figure 2 shows the change in absorption spectrum of zinc tetrabenzoporphyrin at the sequential aza-substitution: absorption bandwidth expansion occurs simultaneously with the increase of the extinction coefficient and bathochromic shift, leading to a decrease in gap width.

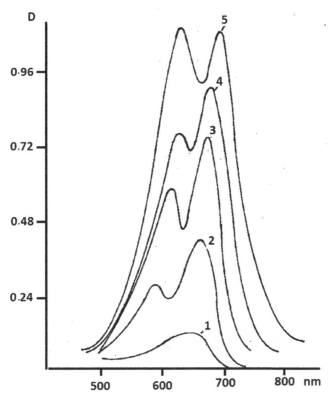

FIGURE 2 The absorption spectra of the films of aza-derivatives of Zn-TBP: 1—Zn–TBP, 2—Zn–MATBP, 3—Zn–DATBP, 4—Zn–TATBP, and 5—Zn–Pc.

Quantitative changes in the quantum yield on a current at sequential aza-substitution are less significant than those for the photocurrent; but nevertheless, the total increase of η reaches 300 percent: the highest increase in efficiency is observed for the first step—a mono-aza-substitution. For the photocurrents, this can be explained by a sharp increase in the extinction coefficient, that is, photochromic properties at the molecular level. In measurements of η, this property has less influence on the final result, and increased quantum yield can be attributed to the high efficiency of energy and charge transfer between the molecules, as well as to a more advantageous arrangement of the band structure energy levels.

Lattice constants in a preferred crystal orientation plane (parallel to the substrate) are 1.98 nm for H_2–Pc and 2.19 nm for H_2–TBP, which corresponds to the axis "a" of the unit cell, connecting centers of molecules lying in one plane. Even on electron micrographs of the film's pigment thickness of 5 nm (Figure 3), obtained with an electron microscope JEM-100B with the ultimate resolution of 0.2 nm at the same scale, there is a clear difference in the interatomic distances and changes in the lattice constants. This difference in interatomic distances and changes in the lattice constants are evident even in electron micrographs of pigment films with a thickness of 5 nm (Figure 3) calculated using an electron microscope JEM-100B with a maximum resolution of 0.2 nm at the same scale.

FIGURE 3 The change in the lattice with fourfold azazameschenii in metal-free tetrabenzporfirin: 1—tetrabenzporfirin and 2—phthalocyanine, resolution of 0.2 nm.

Given the almost identical molecular size (distance between the outer atoms of the benzene rings due to compression of the coordination sphere as a result of aza-substitution in Pc is of 0.13 nm smaller), the minimum distance between the nearest atoms for H_2–Pc is ~0.34 nm and for H_2–TBP is ~0.42 nm. This increases the electron affinity and decreases both the ionization potential and the energy difference between the highest occupied and lowest unoccupied molecular orbitals, leading to a bathochromic shift of the first absorption band of Pc. Simultaneously, there are observed strengthening of the bond to the metal Me^{2+} at the ion radii less than 1.4 A, the appearance of dative π-bonds Me–N, change in the acid–base properties, and in particular, a considerable reduction of protonation even in a strongly acidic medium. Thus, the interatomic distances are reduced by about 20 percent, which contributes to the ring currents in the conjugation macrocycle, to strengthening of intermolecular interactions and increase of the efficiency of energy and charge transfer (tetrapyrrole compounds are characterized by "hopping" charge-transfer mechanism in which intermolecular distances are particularly important) and to an increase of the quantum yield on a current: a maximum value in the investigated group of compounds is 18 percent. Characteristically, the net effect depends on the nature of the central atom, whose interaction with the ligand significantly changes the electron density distribution in the macrocycle and its inner diameter.

Figure 4 shows the position of the energy bands for the limiting cases of substitution (no substitution—TBP, complete replacement–Pc) and the position of the acceptor level of oxygen in accordance with its redox potential.

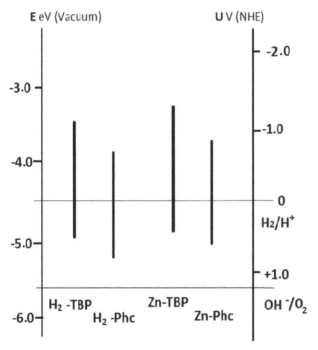

FIGURE 4   Energy band diagram of Me–TBP and Me–Pc.

If the surface layer is filled, that is, oxygen captures an electron to form a charge-transfer complex:

$$O_2 \rightarrow O_{2(a)} \rightarrow [O_2^{-\delta}]_{(a)} + p^{+\delta}$$

then the concentration of adsorbed charged molecules is determined by conventional expression [8]:

$$\vartheta^- = \vartheta\,[1 + 1/2\cdot\exp(w/kT)] = \vartheta\,[1 + 1/2\cdot\exp((E_a - F_s)/kT)]$$

where ϑ—total concentration of adsorbed molecules, w—activation energy of the transition to the charged form, F_s—the distance from the Fermi level at the surface to E_v, and E_a—the distance from the surface acceptor level to the E_v (E_v—the ceiling of the valence band). Consequently, the energy diagrams shown initially determine the increased population of acceptor levels in the dark and weaker photoresponse of TBP compared with Pc.

However, arguments based on the general concepts and physicochemical properties of the molecules do not provide sufficient certainty in assessing the impact of the electronic structure of molecules on the properties of their solid agglomerates. Accordingly, an attempt was made to mathematically model transformation of the structures using quantum chemical calculations of the electron density distribution and energy characteristics of the orbitals.

The calculation of the spatial and electronic structure of molecules was done by the method of molecular orbitals on the program GAUSSIAN 03. In finding the molecular orbitals, there were used density functional theory and the approach of Rutan in which the molecular orbitals are defined in the class of functions of the following form:

$$\varphi_p(\vec{r}) = \sum_{k=1}^{N_{\text{atom}}} \sum_{s=1}^{M(k)} c_{sp}^k \eta_s^k(\vec{r} - \vec{s}_k)$$

where $\eta_1^k(\vec{r}), \eta_2^k(\vec{r}),..., \eta_{M(k)}^k(\vec{r})$—a given set of functions for the k-th atom in the molecule, the position of which is determined by the vector $\vec{s}_k$. These functions, called atomic orbitals, are not assumed to be linearly independent or orthogonal, but are normalized: c_{sp}^k—varying coefficients. The index s is a set of three indices (n, l, m), and the functions $\eta_s^k(\vec{r})$ are

$$\eta_s^k(\vec{r}) = R_{n\ell}^k(r) Y_{\ell m}(\vec{\Omega})$$

Accordingly, the sum over s is a triple sum:

$$\sum_{s=1}^{M(k)} \equiv \sum_{n=1}^{N(k)} \sum_{\ell=0}^{L(k,n)} \sum_{m=-\ell}^{\ell}$$

The set of radial parts of all atomic orbitals used in the construction of the molecular orbitals is the atomic basis of calculation. The basis was taken by a linear combination of Gaussian orbitals:

$$R_{n\ell}(r) \approx \sum_{q=1}^{Q'} d_q^{n\ell} r^{k_q^{n\ell}} e^{-\alpha_q^{n\ell} r^2}$$

in the form of a standard correlation consistent basis set 6-31G**, which is also denoted as *6-31G (1d1pH)* or *cc-pVDZ*. In the calculations, we used two forms of the exchange–correlation functional: form PBEPBE, described in Refs. [9,10], and the form PBE1PBE, described in Refs. [11,12]. To determine the value of the spin in the ground state of the molecule, calculations of molecules with different spins were carried out and the total energies were compared.

The results of calculation by the limited Hartree–Fock method (not taking into account correlation effects) and by the density functional theory (approximately taking into account correlation effects) are summarized in Tables 2 and 3, and the corresponding numbering of atoms is shown in Figure 5.

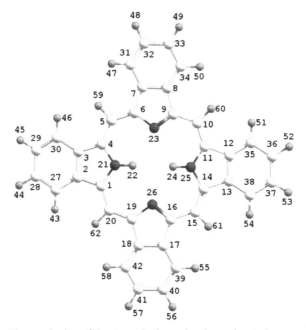

FIGURE 5 The numbering of the atoms in the molecules under study.

TABLE 2 Changes in the distribution of electron density in the molecule of tetrabenzoporphyrin with sequential aza-substitution (Hartree–Fock)

Atom number	Before substitution	0	1	2-ortho	2-para	3	4	After substitution
1	C	0.321051	0.437988	0.274334	0.305926	0.286533	0.584203	C
2	C	−0.08527	−0.04998	−0.03553	−0.0736	−0.06077	−0.10211	C
3	C	−0.08527	−0.12067	−0.14554	−0.11234	−0.12495	−0.1021	C
4	C	0.321052	0.750765	0.654557	0.599145	0.610537	0.584205	C
5	C	−0.16368	−0.73145	−0.68168	−0.64345	−0.65047	−0.63157	N
6	C	0.297132	0.660397	0.638233	0.592743	0.60889	0.588361	C
7	C	−0.01972	−0.12967	−0.06071	−0.05498	−0.04915	−0.04479	C
8	C	−0.10452	−0.01325	−0.10769	−0.09229	−0.11678	−0.11989	C
9	C	0.389694	0.316014	0.688339	0.411316	0.677646	0.664767	C
10	C	−0.24887	−0.17681	−0.74176	−0.24582	−0.71878	−0.70297	N
11	C	0.462781	0.336566	0.76172	0.46861	0.756837	0.746162	C
12	C	−0.07667	−0.09999	−0.13756	−0.06601	−0.10803	−0.0971	C
13	C	−0.07667	−0.07001	−0.03211	−0.10721	−0.0843	−0.0971	C
14	C	0.462781	0.298136	0.397134	0.73886	0.725921	0.746162	C
15	C	−0.24887	−0.14239	−0.18628	−0.71159	−0.68705	−0.70297	N
16	C	0.389694	0.27503	0.335136	0.643082	0.630436	0.664767	C
17	C	−0.10452	−0.01414	−0.10075	−0.13073	−0.13214	−0.11989	C
18	C	−0.01972	−0.10653	−0.0199	−0.01065	−0.00717	−0.04479	C
19	C	0.297132	0.373251	0.247478	0.291599	0.273391	0.588361	C
20	C	−0.16368	−0.2212	−0.11956	−0.15497	−0.13735	−0.63157	N
21	N	−0.80003	−0.86562	−0.78381	−0.78313	−0.78259	−0.76738	N
22	H	0.392221	0.402829	0.400495	0.397033	0.399423	0.402385	H
23	N	−0.77395	−0.77932	−0.80634	−0.78307	−0.78431	−0.77959	N
24	H	0.393755	0.394374	0.409868	0.409103	0.418961	0.425386	H

TABLE 2   *(Continued)*

Atom number	Before substi-tution	0	1	2-ortho	2-para	3	4	After substitu-tion
25	N	−0.88939	−0.7999	−0.84798	−0.87062	−0.85203	−0.85387	N
26	N	−0.77395	−0.76184	−0.75895	−0.77188	−0.76373	−0.77959	N
27	C	−0.0997	−0.12647	−0.12085	−0.10631	−0.11073	−0.09636	C
28	C	−0.16758	−0.14571	−0.14872	−0.16291	−0.15832	−0.16731	C
29	C	−0.16758	−0.16069	−0.18104	−0.1718	−0.17595	−0.16731	C
30	C	−0.0997	−0.09793	−0.08089	−0.09017	−0.08606	−0.09636	C
31	C	−0.13039	−0.10677	−0.11961	−0.11728	−0.12169	−0.1229	C
32	C	−0.14944	−0.16613	−0.14993	−0.14985	−0.14807	−0.14769	C
33	C	−0.16614	−0.1469	−0.15724	−0.15952	−0.15966	−0.16182	C
34	C	−0.11626	−0.13576	−0.11295	−0.12166	−0.11153	−0.11097	C
35	C	−0.1158	−0.09454	−0.09165	−0.12122	−0.10319	−0.10546	C
36	C	−0.1545	−0.17267	−0.16575	−0.14896	−0.15422	−0.15053	C
37	C	−0.1545	−0.16223	−0.14114	−0.15581	−0.14808	−0.15053	C
38	C	−0.1158	−0.10477	−0.12963	−0.10095	−0.10982	−0.10546	C
39	C	−0.11626	−0.13104	−0.1141	−0.10608	−0.10508	−0.11097	C
40	C	−0.16614	−0.14869	−0.17055	−0.16857	−0.17003	−0.16182	C
41	C	−0.14944	−0.16774	−0.15142	−0.14685	−0.1462	−0.14769	C
42	C	−0.13039	−0.11528	−0.12971	−0.13627	−0.1366	−0.1229	C
43	H	0.151604	0.168197	0.151732	0.149808	0.149478	0.167836	H
44	H	0.146236	0.16209	0.148571	0.144667	0.145562	0.143205	H
45	H	0.146236	0.162386	0.14674	0.14434	0.144719	0.143205	H
46	H	0.151604	0.193695	0.1753	0.16955	0.170926	0.167835	H
47	H	0.156107	0.178144	0.185657	0.185897	0.185685	0.18475	H
48	H	0.152006	0.150813	0.157821	0.158502	0.158321	0.157229	H
49	H	0.150338	0.151751	0.156449	0.155772	0.156378	0.154787	H

TABLE 2 *(Continued)*

Atom number	Before substitution	0	1	2-ortho	2-para	3	4	After substitution
50	H	0.155859	0.156186	0.183034	0.160131	0.181867	0.179148	H
51	H	0.168911	0.153006	0.193815	0.170583	0.195648	0.197801	H
52	H	0.160866	0.146531	0.162843	0.164413	0.166934	0.16954	H
53	H	0.160866	0.147148	0.163625	0.165534	0.167665	0.16954	H
54	H	0.168911	0.151038	0.169237	0.196189	0.196229	0.197801	H
55	H	0.155859	0.155679	0.149019	0.174765	0.174001	0.179148	H
56	H	0.150338	0.152402	0.146604	0.148928	0.149169	0.154787	H
57	H	0.152006	0.15058	0.148584	0.150422	0.150828	0.157229	H
58	H	0.156107	0.154874	0.151201	0.154902	0.154583	0.18475	H
59	H	0.16157	0.161812	–	–	–	–	–
60	H	0.150037	–	–	0.160002	–	–	–
61	H	0.150037	0.165883	0.164818	–	–	–	–
62	H	0.16157	0.158493	0.168946	0.16471	0.168241	–	–

TABLE 3 Changes in the distribution of electron density in the molecule of tetrabenzoporphyrin in sequential aza-substitution (by density functional theory on the grid)

Atom number	Before substitution	0	1	2-ortho	2-para	3	4	After substitution
1	C	0.318705	0.335082	0.324168	0.322311	0.336346	0.492694	C
2	C	0.079445	0.072616	0.082218	0.080802	0.07099	0.075752	C
3	C	0.064812	0.083751	0.072442	0.075919	0.086511	0.090177	C
4	C	0.34823	0.481363	0.490204	0.489379	0.477113	0.47302	C
5	C	−0.17246	−0.57271	−0.56948	−0.56449	−0.56676	−0.56427	N

TABLE 3 *(Continued)*

Atom number	Before substi-tution	0	1	2-ortho	2-para	3	4	After substi-tution
6	C	0.297463	0.441181	0.443079	0.438523	0.443982	0.442716	C
7	C	0.083174	0.091614	0.100919	0.087589	0.098647	0.105369	C
8	C	0.100962	0.100847	0.100646	0.106798	0.099981	0.094572	C
9	C	0.286769	0.291985	0.443657	0.291054	0.442895	0.443515	C
10	C	−0.18365	−0.1751	−0.56978	−0.17955	−0.56456	−0.56052	N
11	C	0.318659	0.335855	0.489895	0.322308	0.488486	0.492695	C
12	C	0.079431	0.069906	0.072524	0.080799	0.079997	0.075719	C
13	C	0.064842	0.075938	0.081677	0.075952	0.087275	0.090224	C
14	C	0.348227	0.329671	0.324412	0.489344	0.480765	0.472982	C
15	C	−0.17247	−0.1744	−0.17218	−0.56452	−0.56436	−0.56428	N
16	C	0.297455	0.289554	0.289232	0.43853	0.439488	0.442713	C
17	C	0.08318	0.093555	0.09381	0.087571	0.092703	0.105405	C
18	C	0.100975	0.093156	0.094067	0.106829	0.100872	0.094569	C
19	C	0.286773	0.288464	0.289497	0.291053	0.293325	0.443511	C
20	C	−0.18371	−0.16973	−0.17306	−0.17962	−0.17148	−0.56055	N
21	N	−0.65905	−0.62733	−0.62116	−0.62168	−0.6195	−0.59016	N
22	H	0.294134	0.2974	0.300256	0.301075	0.30412	0.310317	H
23	N	−0.68786	−0.67021	−0.64064	−0.66216	−0.64477	−0.64205	N
24	H	0.294135	0.297583	0.300498	0.30107	0.304198	0.310326	H
25	N	−0.65905	−0.65586	−0.62216	−0.62166	−0.58796	−0.59017	N
26	N	−0.68785	−0.68467	−0.68606	−0.66224	−0.66394	−0.64213	N

TABLE 3 *(Continued)*

Atom number	Before substitution	0	1	2-ortho	2-para	3	4	After substitution
27	C	−0.13023	−0.12928	−0.13376	−0.13117	−0.13039	−0.12134	C
28	C	−0.06303	−0.06185	−0.06018	−0.06119	−0.06131	−0.0614	C
29	C	−0.06419	−0.06384	−0.06413	−0.06341	−0.0622	−0.0601	C
30	C	−0.12825	−0.12248	−0.11767	−0.11866	−0.12307	−0.12376	C
31	C	−0.13672	−0.13249	−0.13799	−0.13083	−0.13473	−0.13856	C
32	C	−0.0652	−0.06233	−0.06072	−0.06316	−0.06073	−0.05982	C
33	C	−0.0628	−0.06243	−0.06059	−0.06106	−0.06048	−0.06057	C
34	C	−0.13974	−0.14349	−0.13721	−0.14275	−0.13808	−0.1338	C
35	C	−0.13025	−0.12735	−0.11733	−0.13118	−0.12374	−0.12133	C
36	C	−0.06303	−0.06337	−0.06442	−0.06118	−0.0611	−0.0614	C
37	C	−0.06419	−0.06322	−0.0601	−0.06343	−0.06033	−0.0601	C
38	C	−0.12825	−0.12971	−0.13326	−0.11867	−0.12509	−0.12376	C
39	C	−0.13672	−0.13633	−0.14028	−0.13082	−0.13341	−0.13856	C
40	C	−0.0652	−0.06374	−0.06313	−0.06316	−0.06204	−0.05982	C
41	C	−0.0628	−0.06362	−0.06298	−0.06105	−0.0615	−0.06058	C
42	C	−0.13974	−0.1404	−0.13993	−0.14276	−0.14191	−0.13379	C
43	H	0.065096	0.063859	0.064969	0.065592	0.066454	0.084349	H
44	H	0.064287	0.064673	0.065905	0.066126	0.067747	0.06954	H
45	H	0.064374	0.065828	0.067117	0.067146	0.06879	0.069374	H
46	H	0.065148	0.081908	0.083745	0.082705	0.084124	0.08418	H
47	H	0.056666	0.073646	0.073892	0.074406	0.07482	0.076813	H

TABLE 3 *(Continued)*

Atom number	Before substitution	0	1	2-ortho	2-para	3	4	After substitution
48	H	0.058693	0.059998	0.061032	0.061571	0.062358	0.063934	H
49	H	0.05859	0.058996	0.061	0.060343	0.062486	0.063877	H
50	H	0.056809	0.056007	0.073897	0.05794	0.07516	0.076201	H
51	H	0.065097	0.066542	0.083793	0.065592	0.083283	0.084348	H
52	H	0.064288	0.066032	0.067165	0.066134	0.068048	0.06954	H
53	H	0.064374	0.066174	0.065889	0.067143	0.068044	0.069375	H
54	H	0.065148	0.066017	0.064993	0.082704	0.082491	0.084174	H
55	H	0.056666	0.056817	0.059009	0.074391	0.076035	0.076817	H
56	H	0.058693	0.059954	0.061288	0.061565	0.063045	0.063933	H
57	H	0.05859	0.060174	0.061263	0.06035	0.061879	0.063876	H
58	H	0.056808	0.058152	0.058946	0.057926	0.058215	0.076205	H
59	H	0.064099	0.067717					H
60	H	0.065772			0.070909			H
61	H	0.064098	0.066716	0.070527				H
62	H	0.065775	0.067212	0.070551	0.070928	0.07273		H

In addition, the spectrum of single-particle states and the spatial distribution of electrons in the highest occupied state (HOMO) and the lowest free state (LUMO) were calculated (Figure 6). The spatial distribution of electron states in the HOMO and LUMO is illustrated in Figures 7–10. These figures show the surface level of functions $|\Psi(\vec{r})|^2$, that is, surfaces satisfying the condition $|\Psi(\vec{r})|^2 = \text{constant}$. Here, $\Psi(\vec{r})$ is the one-particle wave function of the electron in the states HOMO and LUMO of the corresponding molecules: constant values are the same for all molecules.

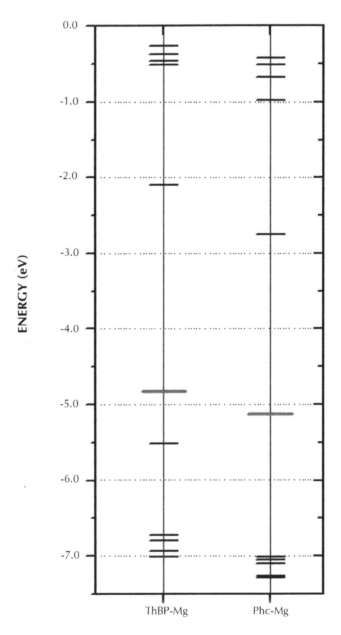

FIGURE 6   Spectrum of one-electron energies in the ground state of molecules Mg-tetrabenzoporphyrin and Mg-phthalocyanine (the spin of the ground state $S = 0$).

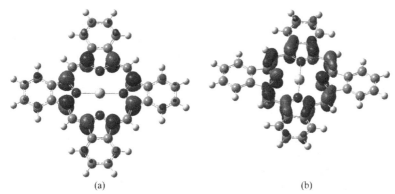

(a)   (b)

FIGURE 7   Spatial distribution of the electron in the one-particle states of HOMO (a) and LUMO (b) for the ground state of Mg-tetrabenzoporphyrin.

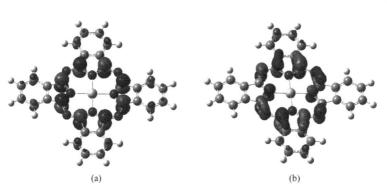

(a)   (b)

FIGURE 8   Spatial distribution of the electron in the one-particle states of HOMO (a) and LUMO (b) for the ground state of Mg-phthalocyanine.

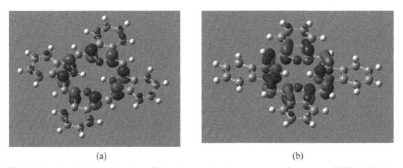

(a)   (b)

FIGURE 9   Spatial distribution of the electron in the one-particle states of HOMO (a) and LUMO (b) for the singlet state of the metal-free tetrabenzoporphyrin.

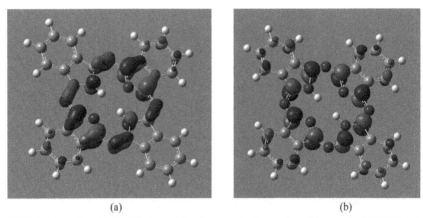

(a)   (b)

FIGURE 10   Spatial distribution of the electron in the one-particle states of HOMO (a) and LUMO (b) for the singlet state of the metal-free phthalocyanine.

In addition, changes of the energy in sequential aza-substitution were assessed (see Figures 11–14).

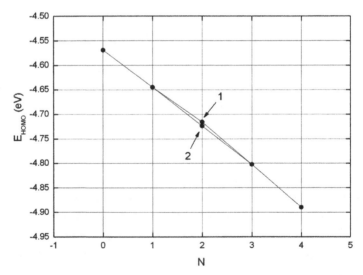

FIGURE 11   Dependence of the energy E_{HOMO} on the number of substitutions in the molecule of tetrabenzoporphyrin.

1—molecule shown in Figure 14(c),

2—molecule shown in Figure 14(d).

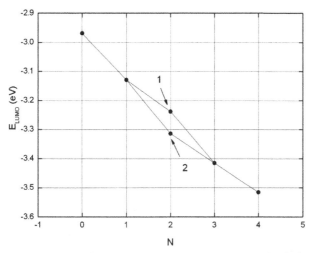

FIGURE 12 Dependence of the energy E_{LUMO} on the number of substitutions in the molecule of tetrabenzoporphyrin.

1—molecule shown in Figure 14(c),

2—molecule shown in Figure 14(d).

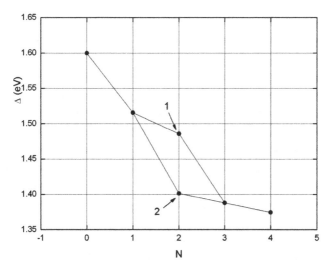

FIGURE 13 Dependence of the energy difference $\Delta E = E_{LUMO} - E_{HOMO}$ on the number of substitutions in the molecule of tetrabenzoporphyrin.

1—molecule shown in Figure 14(c),

2—molecule shown in Figure 14(d).

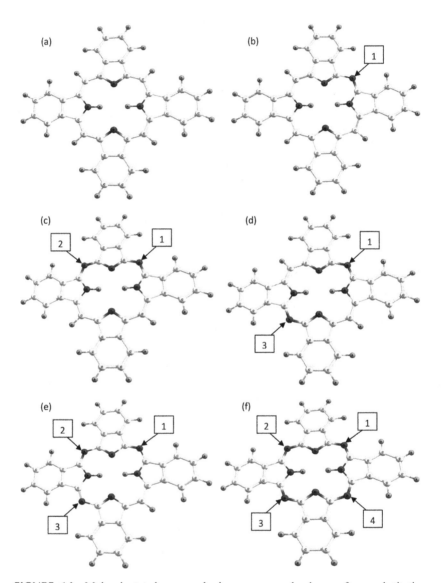

FIGURE 14   Molecule tetrabenzoporphyrin, arrows mark places of aza-substitution (a) tetrabenzoporphyrin (symmetry D2h), (b) mono-substitution (symmetry C1), (c) disubstituted (symmetry D1h), (d) disubstituted (symmetry C2h), (e) triple substitution (symmetry C1), and (f) phthalocyanine (symmetry D2h).

As can be observed from the above calculations, the most revealing chang-es occur, naturally, in the areas of substitution, with the apparent concentration of the charge in the most likely coordination of potential electron acceptor. In similar calculations for other metal derivatives of tetrabenzo-porphyrin, there was no significant difference; however, a significant effect of extraligand in the case of Fe–TBP/FeCl-TBP should be noted. As summarized in Table 1, the ratio of their photocurrents (0.4/1.7 = 0.24) after full aza-substitution (Fe–Pc/ ClFe–Pc) is almost unchanged (1.4/5.4 = 0.26), that is, a significant advantage of the pigment having extraligands on the central atom is saved. The quantum yield of the chlorinated form is five times higher. Therefore, looking forward to the next subject of research, we present some of the data on these com-pounds (see Figures 14–18).

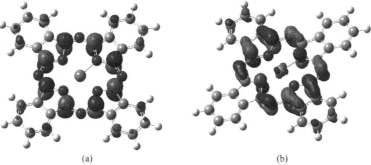

(a)   (b)

FIGURE 15   Spatial distribution of the electron with spin "up" in the one-particle states of HOMO (a) and LUMO (b) for the ground state of Fe-phthalocyanine. Ground-state spin $S = 1$.

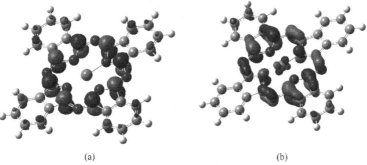

(a)   (b)

FIGURE 16   Spatial distribution of the electron with spin "down" in the one-particle states of HOMO (a) and LUMO (b) for the ground state of Fe-phthalocyanine. Ground-state spin $S = 1$.

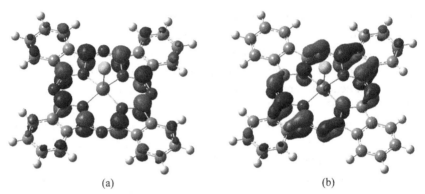

(a)   (b)

FIGURE 17   Spatial distribution of the electron with spin "up" in the one-particle states of HOMO (a) and LUMO (b) for the ground state of FeCl-phthalocyanine. Ground-state spin S = 3/2.

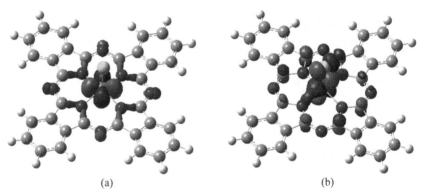

(a)   (b)

FIGURE 18   Spatial distribution of the electron with spin "down" in the one-particle states of HOMO (a) and LUMO (b) for the ground state of FeCl-phthalocyanine. Ground-state spin $S = 3/2$.

The picture of changes in the distribution of electron density is so obvious that practically it does not require any comment. We note only that the most impressive version of Figure 18 is unlikely, as it corresponds to the band gap of about 0.8 eV, and the experimental values (about 1.5 eV) are more consistent with the spin-up states.

2.5 CONCLUSION

Thus, the set of considered measurements on films of aza-substituted metal derivatives of tetrabenzoporphyrin in combination with the data of quantum-chemical calculations of the electron density and energy characteristics of the pigments leads to several conclusions:

- Sequential substitution of carbon atoms in mesoposition by nitrogen that leads to a transition from the porphyrin structure to a phthalocyanine structure increases the electron density in the area under consideration with the corresponding increase of charges in the π-electronic bonds.

- The obvious consequence of the increase of electron density in the important area of structural conjugation of macro-cycle is a ring current increase and compression of the cycle, which leads to a considerable increase of the intermolecular interaction.

- Consequences of the strengthening of the intermolecular interactions are bathochromic red shift of the absorption maxima, the increase in the extinction coefficients, and the broadening of the absorption bands in the visible part of the spectrum, as well as the hypsochromic shift of the Soret band.

- In line with the changes in the optical characteristics of the pigments, a reduction of the band gap takes place, lowering the energy level of the valence band and increasing the depth of acceptor levels formed by the adsorbed oxygen. In this case, it can be seen as a positive change, as it brings pigments to the value of the band gap of 1.5 eV, which is ideal for the converter of solar energy on the ground level.

From a practical viewpoint, the redistribution of electron density in the very structurally similar molecules leads to a significant increase in chemo- and light fastness of pigments and oxidation potentials, to an eight- to tenfold increase in the photocurrent, 1.6- to 1.8-fold increase in photopotentials, and two- to fivefold increase in the quantum yield on a current.

KEYWORDS

- **Aza-substitution**
- **Density distribution**
- **Electron**
- **Organic semiconductors**
- **Porphyrins**
- **Tetrapyrrole compounds**

REFERENCES

1. Rudakov, V. M.; Ilatovsky, V. A.; and Komissarov, G. G.; Photoactivity of metal derivatives of tetraphenylporphyrin. *Khim. Fizika.* **1987**, *6(4),* 552–554 (in Russian).
2. Apresyan, E. S.; Ilatovsky, V. A.; and Komissarov, G. G.; Photoactivity of thin films of metal derivatives of phthalocyanine. *Zh. Fiz. Himii.* **1989**, *63(8),* 2239–2242 (in Russian).
3. Ilatovsky, V. A.; Shaposhnikov, G. P.; Dmitriev, I. B.; Rudakov, V. M.; Zhiltsov, S. L.; and Komissarov, G.G.; Photocatalytic activity of thin films of azasubstituted tetrabenzoporphyrins. *Zh. Fiz. Khimii.* **1999**, *73(12),* 2240–2245 (in Russian).
4. Ilatovsky, V. A.; Dmitriev, I. B.; Kokorin, A. I.; Ptitsyn, G. A.; and Komissarov, G. G.; The influence of the nature of the coordinated metal on the photoelectrochemical activity of thin films of tetrapyrrole compounds. *Khim. Fiz.* **2009**, *28(1),* 89–96 (in Russian).
5. Ilatovsky, V. A.; Ptitsyn, G. A.; and Komissarov, G. G.; Influence of molecular structure of the films of tetrapyrrole compounds on their photoelectrochemical characteristics at the various types of sensitization. *Khim. Fiz.* **2008**, *27(12),* 66–70 (in Russian).
6. Ilatovsky, V. A.; Sinko, G. V.; Ptitsyn, G. A.; and Komissarov, G. G.; Structural Sensitization of Pigmented Films in the Formation of Nano-Sized Monocrystal Clusters. Collection: The Dynamics of Chemical and Biological Processes, XXI Century. Moscow: Institute of Chemical Physics RAS; **2021**, 173–180.
7. Ilatovsky, V. A.; Apresyan, E. S.; and Komissarov, G. G.; Increase in photoactivity of phthalocyanines at structural modification of thin film electrodes. *Zhurn. Fiz. Khimii.* **1988**, *62(6),* 1612–1617 (in Russian).
8. Wolkenstein, F. F.; Physical Chemistry of Semiconductor Surfaces. Moscow: Nauka; **1973**, 398 p.
9. Perdew, J. P.; Burke, K.; and Ernzerhof, M.; *Phys. Rev. Lett.* **1996**, *77,* 3865.
10. Perdew, J. P.; Burke, K.; and Ernzerhof, M.; *Phys. Rev. Lett.* **1997**, *78,* 1396.
11. Ernzerhof, M.; Perdew, J. P.; and Burke, K.; *Int. J. Quantum. Chem.* **1997**, *64,* 285.
12. Ernzerhof, M.; and Scuseria, G. E.; *J. Chem. Phys.* **1999**, *110,* 5029.

CHAPTER 3

POLY(ETHYLENE-*CO*-VINYL ACETATE) COMPOSITES IN NANOSCALE: RESEARCH METHODOLOGY AND DEVELOPMENTS

DHORALI GNANASEKARAN, PEDRO H. MASSINGA JR, and WALTER W. FOCKE

CONTENTS

3.1 INTRODUCTION

There is currently immense interest in the development of nanostructured material (NSMs) for a wide variety of applications, and these materials offer exciting new challenges and opportunities in all the major branches of science and technology.[1] It is widely recognized that reductions in the size of components have an influence on their interfacial interactions, and this factor can, in turn, enhance the material properties to an considerable extent.[2] Consequently, it is also possible to develop materials that are completely discontinuous, that is, which contain both organic and inorganic phases. Such materials exhibit nonlinear changes in properties with respect to composites that are made up of the same phases. This chapter focuses mainly on the classification of NSMs such as zero-dimensional (polyhedral oligomeric silsesquioxane or POSS), one-dimensional (CNT, sepiolite), and two-dimensional (clay) nanostructure materials and on the recent developments in EVA nanocomposites (Figure 1).

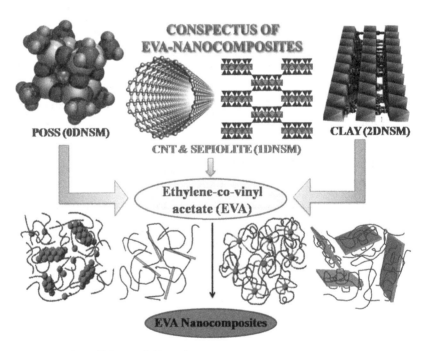

FIGURE 1   Overall theme of this review.

3.1.1 ETHYLENE-CO-VINYL ACETATE

EVA copolymers are one of the most important and widely used organic polymers, with diverse industrial applications, such as electrical insulation,[1] cable jacketing, waterproofing, and corrosion protection, as well as packaging of components,[3] photovoltaic encapsulation, and footwear [4]. EVA is used in paints, adhesives, coatings, textiles, wire and cable compounds, laminated safety glasses, automotive plastic fuel tanks, and so on. It is extremely elastomeric and accepts high filler loadings while retaining its flexible properties. [5] However, bulk EVA does not often fulfill the requirements because of its thermal stability and mechanical properties in some areas. To improve these properties, NSMs are introduced as reinforcing materials. Among several polymeric materials used for polymer nanocomposites, EVA, a copolymer containing repeating units of ethylene as a nonpolar and vinyl acetate (VA) as a polar, has been newly adopted for its polymer NSMs' arrangement because of its potential engineering applications in the fields of packaging films and adhesives.[6]

By changing the VA content, EVA copolymers can be tailored for applications such as rubbers, thermoplastic elastomers, and plastics.[4] The combinations of EVA with NSMs have wide marketable applications, and these applications depend on the VA contained in the main chain. As the VA content increases, the copolymer presents increasing polarity but lower crystallinity, and therefore different mechanical, thermal, and morphological behaviors. The increasing polarity with increasing VA content is apparently useful in imparting a high degree of polymer-NSMs surface interaction, and it has been reported that there is a rise in the Young's modulus and yield strength of EVA polymeric nanocomposites [7] compared with other polymeric nanocomposites.

3.1.2 CLASSIFICATION OF NSMs

The commencement of research into nanotechnology can be traced back to over 40 years; but in the past decade, hundreds of NSMs have been obtained across a variety of disciplines. NSMs are low-dimensional materials comprising of building units of submicron or nanoscale size in at least one direction and exhibiting size effects.[8]

Shape of nanostructure materials

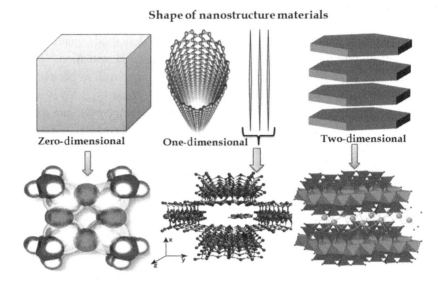

FIGURE 2   Chemical structure and shape of nanostructured materials.

The first part of this review can be classified into the three different types according to the dimensionality of nanomaterials (Figure 2): (1) 0DNSMs, such as POSS, which are characterized by three nanometric dimensions; (2) 1DNSMs (fibrous materials), such as CNTs and sepiolite, which are characterized by elongated structures with two nanometric dimensions; and (3) 2DNSMs (layered materials), such as clay (e.g., montmorillonite: MMT), which are characterized by one nanometric dimension.

1.3   ZERO-DIMENSIONAL NANOSTRUCTURED MATERIALS

Zero-dimensional nanostructured materials (0DNSMs) are those in which all the dimensions are measured within the nanoscale (no dimensions, or zero-dimensional, are larger than 100 nm). The most common representation of 0DNSMs is nanoparticles.

Owing to their large specific surface area and other properties superior to those of their bulk counterparts arising from the quantum size effect, 0DNSMs have attracted considerable research interest and many of them have been synthesized in the past 10 years.[9–11] It is well known that the behaviors of

NSMs depend strongly on their sizes, shapes, dimensionality, and morphologies, which are thus the key factors giving rise to their ultimate performance and applications. It is therefore of great interest to synthesize 0DNSMs with a controlled structure and morphology. In addition, 0DNSMs are important materials due to their wide range of applications in the areas of catalysis, as magnetic materials and as electrode materials for batteries.[12] Moreover, the 0DNSMs have recently attracted intensive research interest because the nanostructures have a larger surface area and supply enough absorption sites for all involved molecules in a small space.[13] In addition, the property of porosity in three dimensions could lead to better transport of molecules.[14]

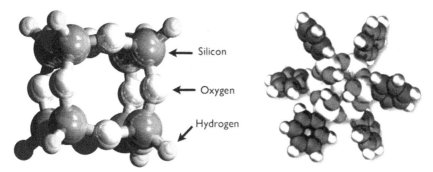

FIGURE 3   Chemical structure of polyhedral oligomeric silsesquioxane.

POSS is a class of organic–inorganic hybrid 0D nanostructure material constituted of an inorganic silica,[15] which consists of a rigid, crystalline silica-like core, having the general formula $(RSiO_{1.5})_a(H_2O)_{0.5b}$, where R is a hydrogen atom or an organic group and a and b are integers ($a = 1, 2, 3,\ldots$; $b = 1, 2, 3,\ldots$), with $a + b = 2n$, where n is an integer ($n = 1, 2, 3,\ldots$) and $b \leq a + 2$. POSS is unique with regard to size (0.5 nm in core diameter) when compared with other nanofillers and can be functionally tailored to incorporate a wide range of reactive groups.[16] The size of a POSS molecule is nearly 1.5 nm in diameter and about 1,000 D in mass; hence, POSS nanostructures are approximately equivalent in size to most polymer dimensions and smaller than the polymer radii of gyration. Figure 3 shows the chemical skeleton of one of the POSS. POSS systems may be viewed as the smallest chemically discrete particles of silica possible, whereas the resins in which they are incorporated may be viewed as nanocomposites, which are intermediate between polymers and ceramics.

POSS derivatives have two unique features: (1) the chemical composition of POSS ($RSiO_{1.5}$) was found to be intermediate between that of silica (SiO_2) and siloxane (R_2SiO), and (2) POSS compounds can be tailored to have various functional groups or solubilizing substituents that can be attached to the POSS skeleton.

One of the most popular branches of silsesquioxanes is POSSs, including the T_8 cage, T_{10} cage, T_{12} cage, and other partial cage structures. Cubic structural compounds (completely and incompletely condensed silsesquioxanes) are commonly illustrated as T_6, T_7, T_8, T_{10}, and T_{12} based on the number of silicon atoms present in the cubic structure (Figure 4). The silica core of POSS is inert and rigid, whereas the surrounding organic groups provide compatibility with the matrix and processability. However, much more attention has been directed to the silsesquioxanes with specific cage structures designated by the abbreviation POSS.[17] Kudo et al. [18] explored various stages of the most plausible mechanism for the synthesis of POSS, and an entire reaction scheme, including all intermediates, was considered.

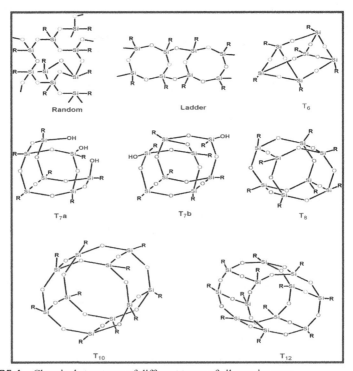

FIGURE 4   Chemical structures of different types of silsesquioxanes.

Conceptually, POSS may be thought of as an organic–inorganic hybrid (Figure 5). Similarly, POSS is sometimes considered to be a filler and sometimes a molecule. For example, POSS is rigid and inert like inorganic fillers; but unlike those conventional fillers, POSS can dissolve molecularly in a polymer.

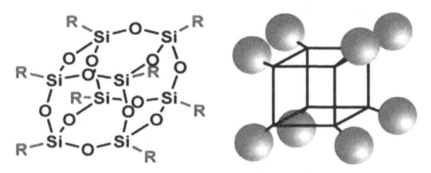

FIGURE 5   Silsesquioxanes Q_8 ($Q = SiO_{2/2}$); R = H, vinyl, epoxy, acetylene, and acrylate.

3.1.4 ONE-DIMENSIONAL NANOSTRUCTURED MATERIALS

Within the various branches of nanotechnology, one-dimensional nanostructures (1DNSMs) have paved the way for numerous advances in both fundamental and applied sciences. 1D NSMs have one dimension that is outside the nanoscale. This leads to needle-shaped nanomaterials. One-dimensional materials, which include nanotubes, nanorods, and nanowires, with at least one dimension in nanometer size fall under the category of 1DNSM. Almost all classes of materials, that is, metals, semiconductors, ceramics, and organic materials, have been used to produce 1DNSMs. However, CNTs (Figure 6) have occupied a most significant place and are the most widely studied 1DNSM.[19]

It is generally accepted that 1DNSMs are ideal systems for exploring a large number of novel phenomena at the nanoscale and investigating the size and dimensionality dependence of functional properties. Certain fields of 1DNSMs, such as nanotubes and sepiolite, have received significant attention.

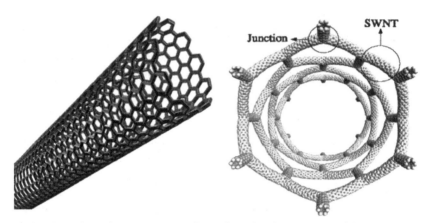

FIGURE 6   Schematic representation of one-dimensional nanostructured CNT.

1DNSMs have had a profound impact on nanoelectronics, nanodevices, and nanocomposite materials. CNTs are long, slender fibers formed into tubes. The walls of the tubes are hexagonal carbon (as shown in Figure 6), and the tubes are often capped at each end.[20] CNTs have been found to be effective reinforcing agents for several polymeric materials, apart from their ability to increase the electrical and thermal conductivity of these materials.[21,22] Since Chaudhary et al. [23] first reported their existence, they have attracted increasing attention because of their high electrical and thermal conductivity, mechanical strength, and chemical stability.[24]

Sepiolite is a family of fibrous hydrated magnesium silicates with the theoretical half unit-cell formula $Si_{12}O_{30}Mg_8(OH)_4(OH_2)_4 \cdot 8H_2O$.[25] The chemical structure of sepiolite shown in Figure 7 is similar to that of the 2:1 layered structure of MMT, consisting of two tetrahedral silica sheets enclosing a central sheet of octahedral magnesia, except that the layers lack continuous octahedral sheets.[26] The discontinuity of the silica sheets gives rise to the presence of silanol groups (Si–OH) at the edges of the tunnels, which are the channels opened to the external surface of the sepiolite particles.[27] The presence of silanol groups (Si–OH) can enhance interfacial interaction between sepiolite and polar polymers.[28]

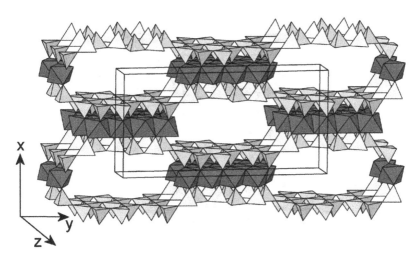

FIGURE 7   Schematic representation of one-dimensional nanostructured sepiolite.

3.1.5 TWO-DIMENSIONAL NSMs WITH ETHYLENE-CO-VINYL ACETATE

Two-dimensional nanomaterials do not have two of the dimensions confined to the nanoscale, that is, 2D nanostructured materials (2DNSMs) have two dimensions outside of the nanometric size. 2DNSMs exhibit plate-like shapes and include nanolayers and nanocoatings.

In recent years, 2DNSMs have become a focus area in materials research, owing to their many low-dimensional characteristics, which differ from the bulk properties.[29] In the quest for 2DNSMs, considerable research attention has been focused over the past few years on the development of 2DNSMs. 2DNSMs with certain geometries exhibit unique shape-dependent characteristics, and they are consequently utilized as building blocks and key components of nanodevices.[30–32] The 2DNSM of clay is shown in Figure 8.

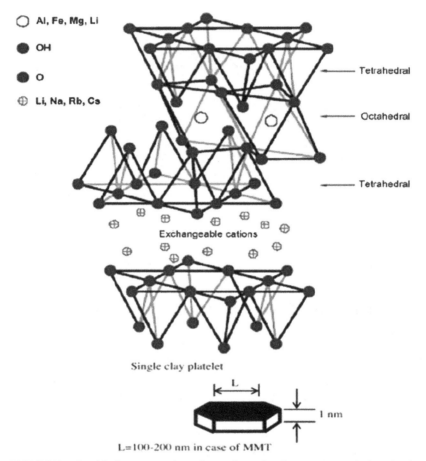

FIGURE 8   Graphical representation of two-dimensional nanostructured clay platelet. [33]

3.2   POLYMER NANOCOMPOSITES

Nanoscale fillers, which are considered to be very important, include layered silicates (such as MMT), nanotubes (mainly CNTs), fullerenes, SiO_2, metal oxides (e.g., TiO_2, Fe_2O_3, Al_2O_3), nanoparticles of metals (e.g. Au, Ag), POSS, semiconductors (e.g., PbS, CdS), carbon black, nanodiamonds, etc. Clay systems have been well developed, followed by POSS; no much development has occurred using graphite–polymer and nanotube–polymer nanocomposites. In addition to clays, nanocomposites have been prepared using POSSs, graphites, and CNTs. The interest in such systems (organic–inorganic hybrid materials) is because the ultrafine or nanodispersion of filler, as well as the local interactions between the matrix and the filler, leads to a higher level of properties than for equivalent micro- and macrocomposites.[34] In a nanocomposite, the clay, or the nanofiller/additive, is well dispersed throughout the polymer. Polymer–clay nanocomposites are a new class of composite materials consisting of a polymer matrix with dispersed clay nanoparticles. In recent years, more attention has been given to incorporating nanomaterials into polymer matrices to obtain high-performance nanocomposites. Clay typically consists of particles with a high aspect ratio, that is, the length of the particle is much longer than their width. Dispersion of the filler on a nanometer scale generally gives the polymer interesting insulation properties.

3.2.1 POSS NANOCOMPOSITES

Due to the great chemical flexibility of POSS molecules, POSS can be incorporated into polymers by copolymerization, grafting, or even blending using traditional processing methods, and it can lead to a successful improvement of the flammability and thermal or polymer mechanical properties.[2,35] Unlike traditional organic compounds, most POSS compounds release no volatile organic compounds below 300°C, are odorless, nontoxic, and reduce the toxicity of smoke upon combustion. Apart from this, the problems associated with polymer immiscibility are reduced. In the polymer–POSS nanocomposites, the POSS acts as a nanoscale-building block, and its interaction with the polymers on a molecular level is considered to be helpful for efficient reinforcement. Polymer–POSS nanocomposites are defined as polymers having small amounts of nanometer-size fillers (POSS), which are homogeneously dispersed by only several weight percentages (Figure 9). A polymer–POSS nanocomposite with a filler having a small size leads to a dramatic increase in the interfacial area compared with traditional composites. This interfacial area creates a significant volume fraction of interfacial polymer with properties different from those of bulk polymer even at low loadings.[36,37]

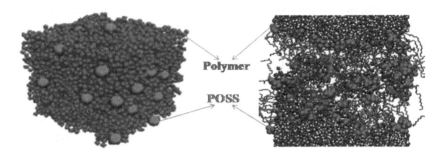

FIGURE 9   Systematic representation of polymer–POSS nanocomposites.

Normal fillers and especially nanofillers suffer from agglomeration. The agglomerates formed when conventional fillers are used lead to weak points in the polymer (stress concentrations), and this leads to poor impact resistance and poor elongation, thereby it will break. As a molecule, POSS dissolves in polymers as 1–3 nm cages and this gives performance advantages not seen with fillers. POSS increases the modulus of elastomers due to the stiffness of the cage and the high cross-link densities attainable using polyfunctional POSS cross-linkers. These include increased modulus and strength, outstanding barrier properties, improved solvent and heat resistance, and decreased flammability. This nature allows them to exhibit properties different from those of conventional microcomposites.

Organic–inorganic composite materials have been extensively studied for a long time. These may consist of two or more organic–inorganic phases in some combined form with the constraint that at least one of the phases or features must be nanosized. Such materials combine the advantages of the organic polymer (e.g., flexible, dielectric, ductile, and processable) and those of the inorganic material (e.g., rigid and thermally stable). Because of these reasons, nanocomposites find use in new applications in many fields such as for gas separation, [8(a),38] in the aerospace industry,[39] for electrical applications,[40] as mechanically reinforced lightweight components, and in nonlinear optics, solid-state ionics, nanowires, and sensors.

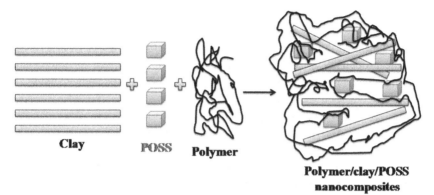

Clay   POSS   Polymer

Polymer/clay/POSS
nanocomposites

FIGURE 10   Formation mechanism of polymer/clay/POSS nanocomposites.

Fox et al. [41] prepared a POSS-tethered imidazolium surfactant and used it to exchange MMT for the preparation of polymer nanocomposites in poly(ethylene-*co*-vinyl acetate, EVA) using a melt-blending technique. Vaia et al. [42] suggested that the extent of exfoliation of clay layers in a melt-blended polymer composite is a result of two competing factors: enthalpic loss due to unfavorable polymer clay interactions and entropic gain due to increased polymer mobility over confinement within an intercalated clay structure. In addition, the presence of POSS interactions add to the enthalpic barrier to exfoliation, as the POSS crystal domains must be melted. Despite these barriers to exfoliation, polymer composites exhibit small increases in exfoliation and smaller tactoids over unmodified clay. It is conceivable that the more rigid structure of POSS-imidazolium relative to other organic modifiers creates a more permanent barrier between the charged clay surface and more hydrophobic polymer chain. The model reaction of POSS/clay composite is as shown in Figure 10.

Fox et al. [43] reported the synthesis and characterization of 1,2-dimethyl-3-(benzyl ethyl iso-butyl polyhedral oligomeric silsesquioxane) imidazolium chloride (DMIPOSS-Cl) and DMIPOSS-modified MMT (DMIP-MT) at several loadings of DMIPOSS-Cl and investigated the ability of these clays to exfoliate in poly(ethylene-*co*-vinyl acetate) systems. In addition, they reported the effects of partial clay loading using only the POSS-modified imidazolium surfactant on the extent of exfoliation and quality of dispersion in polymer. In addition, Zhao et al. [44] showed that although the organic groups on POSS do lower the polarity of clay surface, the surface free energy of clay exchanged with aminopropylisooctyl POSS is reduced by only half, leaving the clay surface still significantly more polar than polymer. As clay is only partially exfoliated upon incorporation of polymers, regardless of their polarity, it is likely that the propensity for POSS to aggregate and limited space for intercalate polymer chains to move dominate the behavior in POSS-exchanged clays.

3.2.2   CNT AND SEPIOLITE NANOCOMPOSITES

The one-dimensional nanostructure of CNTs, their low density, their high aspect ratio, and extraordinary properties make them particularly attractive as reinforcements in composite materials. These studies have been discussed in some excellent reviews.[45,21] The variation of many parameters, such as CNT type, growth method, chemical pretreatment as well as polymer type, and processing strategy, has given some encouraging results in fabricating relatively strong CNT nanocomposites. Since the early preparation of a CNT/epoxy composite by Ajayan et al. [46], more than 30 polymer matrices have been investigated with respect to reinforcement by CNTs. The outstanding potential of CNTs as reinforcements in polymer composites is evident from the super tough composite fibers fabricated by Baughman et al. [24] Till now, hundreds of publications have reported certain aspects of the mechanical enhancement of different polymer systems by CNTs.

CNTs have clearly demonstrated their capability as fillers in diverse multifunctional nanocomposites. The observation of an enhancement of electrical conductivity by several orders of magnitude at very low percolation thresholds (< 0.1 wt %) of CNTs in polymer matrices without compromising other performance aspects of the polymers, such as their low weight, optical clarity, and low melt viscosities, has triggered an enormous activity worldwide in this scientific area. Nanotube-filled polymers could potentially, among others, be used for transparent conductive coatings, electrostatic dissipation, electrostatic painting, and electromagnetic interference shielding applications. A wide range of values for conductivity and percolation thresholds of CNT composites have been reported in the literature during the past decade, depending on the processing method, polymer matrix, and type of nanotube.

A majority of research in polymer/clay nanocomposites is focused on platelet-like clays, smectite clays such as MMT, but only very few works have been focused on fiber-like clays particles (sepiolite).[47] Because of the peculiar shape, these nanofillers are believed to be good candidates for the preparation of nanocomposite materials. In fact, the dispersion of needle-like clays, compared with platelet-like clays, is favored by the relatively small contact surface area. Moreover, the reinforcement capacity of fibers in polymer nanocomposites is higher than that of platelet for uniaxial composites. In this research, polymer nanocomposites with different types of polymer matrices have been taken into consideration: polypropylene (PP), polyamide 6 (PA6),[48] polyurethane,[49] and acrylonitrile–butadiene–styrene.[50]

3.2.3   CLAY NANOCOMPOSITES

Clays have been widely investigated and used as reinforcing agents for polymer matrices.[21,51–53] They can dramatically enhance the mechanical performance and the barrier properties at filler loadings as low as 3–5 wt %, without significantly changing other important characteristics such as transparency or density. To enhance their compatibility with polymer matrices, clays are usually modified with organic compounds (e.g., quaternary ammonium salts), but, even in this case, the properties of the polymer matrices are often not significantly improved.[51] Various researchers have reported on properties of nanoclay–EVA nanocomposites.[54,55]

IUPAC defines a composite as "a multicomponent material comprising multiple different (nongaseous) phase domains in which at least one is a continuous phase." Polymer–clay composites can be prepared in three different ways, namely, by *in situ* polymerization, [56,57] by solution intercalation, and by dispersion through melt processing. Three organoclay nanocomposite microstructures are possible, that is, phase separated, intercalated, and exfoliated. The outcome is determined by which one of the interfacial interactions is favored in the system. The three main interactions are polymer–surface, polymer–surfactant, and surfactant–surface.[58] The consensus is that, for thorough clay sheet dispersion, highly favorable polymer–surface interactions are essential.[58] A greater degree of exfoliation and better dispersion of layered double hydroxides (LDHs) were obtained in more polar matrices. Polyolefin nanocomposites are difficult to prepare as the low polymer polarity does not provide effective interaction with LDHs. Addition of the maleic anhydride-grafted polyethylene as compatibilizer improves, to some extent, the dispersion of clays in nonpolar matrices during melt compounding.[59] The dispersed form of clay platelets is shown in Figure 11. It is well established that the dispersion of particles with high aspect ratios, such as fibers and platelets, in polymeric matrices improves mechanical stiffness as well as some other properties.

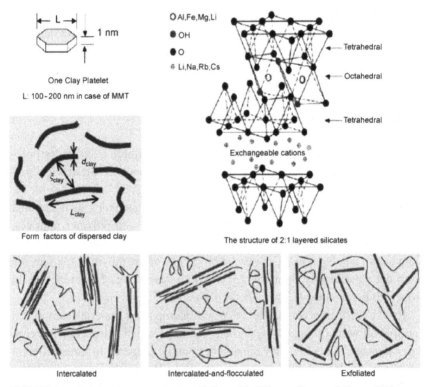

FIGURE 11   Schematic representation of clay with different dispersed phases.[51]

However, good interfacial adhesion and a homogeneous dispersion are prerequisites.[60] Nanostructured clays are ideal for the preparation of polymer–clay nanocomposites that possess improved gas barrier properties, better mechanical properties,[61] enhanced flame retardancy,[62] UV stabilization,[63] or enhanced photodegradation,[64] and so forth.

3.3   EVA NANOCOMPOSITES
3.3.1   POSS–EVA NANOCOMPOSITES

EVA represents a considerable portion of the polymers currently marketed and has a wide range of applications, for example, in nonscratch films, hoses, coatings, and adhesives.[65] It is commonly used in blends with polyolefins in order to improve the mechanical strength, processability, and impact strength and insulation properties. The demands for specific polymer properties for new applications have increased, including their use at higher temperatures and greater resistance to oxidation. The polymer industry has been able to keep abreast of these market demands through the use of additives, fillers, and polymer blends. More recently, the diverse and entirely new chemical technology of POSS nanostructured polymers has been developed.[16,66] This technology affords the possibility of preparing plastics that contain nanoscale reinforcements directly bound to the polymer chains.

The incorporation of POSS cages into polymer materials may result in improvements in polymer properties, including increased temperature of usage, oxidative resistance, and surface hardening, resulting in better mechanical properties as well as a reduction in flammability and heat evolution.[67] In nanocomposites, the addition of POSS nanoparticles can lead to the thermal stabilization of polymers during their decomposition.[68] However, in this regard, the effect of the nanoparticle content is very crucial. In most cases, the thermal stability enhancement takes place at a low loading (4–5 wt %) of POSS nanoparticles; whereas at high loading, thermal stabilization becomes progressively lower. This is because at higher concentrations nanoparticles can form aggregates, and thus the effective area of nanoparticles in contact with polymer macromolecules is lower. In this case, microcomposites may be formed instead of nanocomposites, and thus the protective effect of nanoparticles becomes lower.

In EVA systems, the presence of 1 wt % of POSS slightly increases the values of the stabilized and totalized torque and specific energy. This suggests that the viscosity is higher, probably due to an increase in the extent of entanglement caused by the incorporation of POSS into the polymer matrix.[69] A decrease in these values for a POSS content of 5 wt % was also observed, suggesting that, in high concentrations, POSS forms aggregates, which are not incorporated into the polymer, leading to a decrease in viscosity. However, for EVA systems, there is an increase in the stabilized and totalized torque and in the specific energy with an increase in the POSS content.

X-ray diffraction (XRD) analysis was carried out to verify the crystal structure of the POSS and of the polymer matrix. The XRD diffractograms of the POSS and the systems under study are shown in Figure 12. For pure POSS, Figure 12 shows peaks at $2\theta = 8°$, 8.8°, 10.9∘, 11.7°, 18.3°, and 19.7°, which are characteristic of its crystalline structure. The EVA copolymer (Figure 12, 0 wt % of POSS) shows broad peaks (2θ ranging from 21° to 25°) related to the polyethylene structure of the EVA copolymer. For the other systems (Figure 12), these peaks appear for POSS concentrations of 1 wt % and above, indicating that the presence of EVA in the composite leads to the appearance of POSS aggregates at lower concentrations, probably due to the increase in the polarity of the polymer matrix.

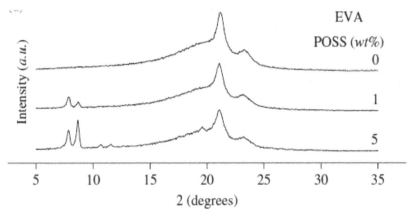

FIGURE 12   XRD patterns of POSS and of the composites.[70]

The morphology and thermal properties of EVA composites with POSS nanostructures were analyzed and compared with those of the pure polymers. The POSS underwent aggregation at higher concentrations during composite processing, indicating a solubility limit of around 1 wt %. The presence of EVA in the composite favors POSS aggregation due to the increase in the polarity of the polymer. These aggregates were observed in Si mapping and were characterized by the presence of a melt peak of POSS.

Further, these aggregates indicate that the polarity of the polymeric matrix plays a major role in the composite's morphology, even for immiscible systems. Figures 13 and 14 show SEM and Si mapping of 1 and 5 percent of POSS in EVA nanocomposites, respectively.

FIGURE 13   SEM (*left*) and Si mapping (*right*) of the composites with 1 wt % of POSS.[70]

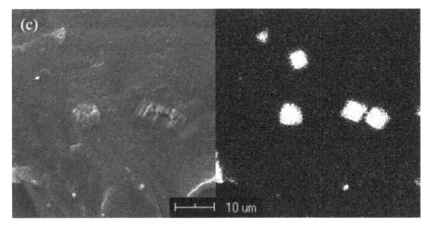

FIGURE 14   SEM (*left*) and Si mapping (*right*) of the composites with 5 wt % of POSS.
[70]

The differential scanning calorimetry (DSC) curves of the second heating for the systems are shown in Figure 15. EVA has Tm values of 84°C. The EVA systems with 5 wt % of POSS also show a third melt peak, at approximately 56°C, which is characteristic of POSS. In the system with 1 wt %, the POSS appears to be homogenously dispersed in the polymer matrix, minimizing the size of the domains in the samples. However, at a concentration of 5 wt %, aggregation of the POSS takes place, which is characterized by the melt peak.

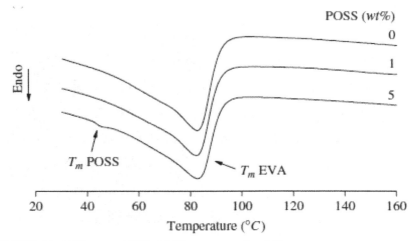

FIGURE 15   DSC curve of EVA–POSS nanocomposites.[70]

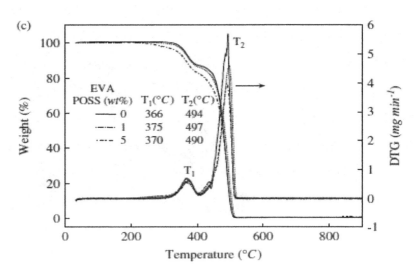

FIGURE 16   TGA curves of EVA–POSS nanocomposites.[70]

The thermogravimetric analysis curves of the systems are shown in Figure 16. The first weight loss (WL) is observed in the temperature (T_1) range of 356–374°C, which is related to the elimination of the acetic acid in EVA and the degradation of POSS. In all systems, as the POSS content increases, the values of the onset temperature for degradation decrease due to the lower

POSS thermal degradation temperature (285°C). Fina et al. [66] obtained similar results for the thermal degradation of PP/POSS nanocomposites.

3.3.2 CNT–EVA AND SEPIOLITE–EVA NANOCOMPOSITES

The CNT is one of the stiffest materials produced commercially, having excellent mechanical, electrical, and thermal properties. The reinforcement of rubbery matrices by CNTs was studied for EVA. George et al. [71] investigated the tensile strength of EVA-CNT and showed that it increased greatly (to 61%), even for very low fiber content (1.0 wt %). The introduction of even a small number of CNTs can lead to improved performance of EVA. At 4 wt % CNT loading, both the modulus and the tensile strength of the nanocomposite increased substantially. However, similar improvements were not observed at higher (8 wt %) nanofiller loading due to filler agglomeration. George and Bhowmick [3] explained the effect of nanofillers and the VA content on the thermal, mechanical, and conductivity properties of nanocomposites. They showed that polymers with high VA content have more affinity toward fillers due to the large free volume available, which allows easy dispersion of the nanofillers in amorphous rubbery phase, as confirmed from morphological studies. The thermal stability of nanocomposites is influenced by type of nanofiller.

Beyer [72] studied the flame-retardant EVA-CNT nanocomposites and was synthesized by melt-blending. Fire property measurements using cone calorimeter revealed that the incorporation of CNT into EVA significantly reduced the peak heat release rates compared with the virgin EVA. Peak heat release rates of EVA with CNT were slightly improved compared with EVA nanocomposites based on modified layered silicates. There was also a synergistic effect by the combination of CNTs and organoclays resulting in an overall more perfect closed surface with improved heat release values. CNTs are highly effective flame retardants; they can also be more effective than organoclays. There was a synergistic effect between organoclays and nanotubes; the char formed during the degradation of the compound by a cone calorimeter was much less cracked compared with those from the single organoclay or CNT-based compounds.

Huang et al. [56] prepared EVA–sepiolite composites and observed a remarkable change in the Young's modulus when only a small amount of sepiolite was incorporated. The fact that sepiolite has little effect on the elongation at break of the EVA/MH/SP composites, but a distinct effect on the Young's modulus, can probably be ascribed to the complex interactions between the polar EVA and the silanol groups of the high aspect ratio sepiolite. Hydrogen bonding may also be expected to occur between the ester groups of EVA and the characteristic silanol groups.

3.3.3   CLAY-EVA NANOCOMPOSITES

Melt-intercalated nanocomposites of EVA with layered silicates have been studied over the past decade.[4,6,8(b),57,73–81] The method is an alternative to *in situ* polymerization and solution intercalation in the preparation of polymer nanocomposites. It is the most environment friendly method, versatile and compatible with current processing equipment.

Melt intercalation consists of blending molten polymer matrices with silicates. Polymer chains may diffuse from the matrix into the silicate interlayer. Conventionally, polymer chains should first intercalate into the silicate interlayers. Subsequently, the chains may push the silicate platelets further apart. Individual silicate particles may thus exfoliate into the polymer matrix. Sufficient compatibility between the surface energy of the silicate layers with that of the polymer chains is required. Silicate layers are hydrophilic, tending to be compatible only with polar polymers. Nonpolar polymers appear to exhibit interfacial adhesion with organosilicates. However, compatibility between the polymer matrix and the surfactant chains of organosilicates is indispensable. It maximizes the freedom of the organosilicate chains in the polymer matrix configurationally. Hence, organosilicate chains gain entropy, which balances entropy loss owing to polymer chain confinement. It has been shown that only polar polymer chains are attached to silicate platelets, whereas nonpolar chains protrude into the bulk melt. Therefore, the interfacial adhesion with organosilicates is insufficient to yield thermodynamically stable nanocomposites.

Delaminating silicate platelets and achieving exfoliation requires strong shear forces.[82] These are attained during nanocomposite processing. Exfoliated structures are those in which the clay platelets are delaminated and individually dispersed in the polymer matrix. In exfoliated nanocomposite structures, silicate platelets are extensively interspaced. Such extensive interlayer separations disrupt coherent layer stacking, and the resultant ordering of the clay platelets is not sufficient to produce a scattering peak. Hence, a featureless XRD diffraction pattern is recorded. Conversely, intercalated nanocomposites are those in which polymer chains and silicate layers alternate, in a well-ordered multilayer arrangement, with a well-defined and well-preserved interlayer distance.

Exfoliation is a key requirement for improving polymer properties. These include mechanical, thermal, barrier, and flame-retardant and optical properties. Exfoliation can lead to a very large surface area for the interaction of stiff silicate particles with polymer chains. EVA, $-(CH_2CH_2)_n[CH_2CH(OCOCH_3)]_m-$, is a somewhat polar copolymer due to its VA units. A higher content of VA, $-CH_2CH(OCOCH_3)-$, affords polarity for interaction with (OH groups of) pristine silicates.[77(a)] Good surface affinity between EVA and silicates alone is not enough to achieve exfoliation—prior break-up and expansion of intrinsic silicate stacks is required. This is accomplished through organic intercalation, using ion-exchange reactions. The stacks intrinsic to silicates are explained as follows: layered silicates comprise particles with thicknesses in the nanometer range; their aspect ratio is very large, exceeding 200; there is a propensity for aggregation of such particles due to the large packing area.

Generally, mechanical shear forces break-up intrinsic silicate stacks. Subsequently, intercalation of large organic species expands the silicate interlayers. This expansion facilitates the diffusion of polymer chains into the layers. This, in turn, promotes the exfoliation of silicate platelets in polymer matrices. Ultimately, a large interfacial area for interaction between polymer chains and silicate particles is thus created. Intercalation may further alter the silicate surface chemistry. The changing nature of the silicate layers from hydrophilic to organophilic renders them compatible with hydrophobic polymers. To date, the preparation of fully exfoliated EVA-layered silicate nanocomposites remains a challenge. The use of different types of processing equipments and varying conditions has yielded disparities between nanocomposite structures and properties. This has resulted from the use of EVA with diverse features, silicates and organic modifiers of different types, modifiers with several chain lengths and different polarities, all intercalated into silicates in various amounts. It is therefore necessary to understand the influence of materials, instruments, and processing conditions on the properties of created composites.

The transmission electron microscopy (TEM) shows the structures of the composites. Structures probed solely using XRD analysis are not completely certain. When no XRD peak is observed, it could be deduced that exfoliation has taken place, whereas the presence of a small number of ordered stacks may also yield a featureless XRD diffractogram. Moreover, the presence of a high number of nonuniformly dispersed clay stacks likewise yields a featureless pattern or interlayer variations. In turn, the presence of any homogeneous distribution of the clay nanoplatelets could be interpreted as intercalation. Further, the presence of few silicate stacks could be deemed to indicate conventional microcomposites. Certainly, the presence of recalcitrant XRD reflections confirms the occurrence of intercalated or unmodified silicate regions. However, exfoliated regions may perhaps coexist and even predominate. TEM is critical as it shows direct evidence of composite structures. However, it may also fall short when a few samples not representative of the whole material are used. The research work reviewed below has probably been subject to this dilemma.

3.4   EFFECT OF EQUIPMENT, TYPES OF CLAY, AND EVA ON NANOCOMPOSITES

3.4.1   EFFECT OF PROCESSING EQUIPMENT AND CONDITIONS ON DISPERSION OF EVA NANOCOMPOSITES

A co-rotating twin-screw extruder is considered the most effective shear device for the dispersion of silicate platelets. This is because the screws rotate in the same direction, intermesh, and pass resin over and under one screw to another. The material is thus subjected to identical amounts of shear and unlikely to become stagnant. Chaudhary et al. [23] and Pistor et al. [79(b)] reported similar EVA nanocomposite structures. Partial intercalation and exfoliation were the predominant characteristics reported in both communications. It was further stated that all the samples contained small tactoid fractions.[23,79(b)] However, the processing equipment and conditions employed by these researchers were different.

La Mantia et al. [78(a)] prepared equivalent samples using two different extruders. They reported similar EVA nanocomposite structures comprising intercalated silicate domains, as detected by XRD. The d_{001} was slightly increased for samples compounded in a twin-screw extruder.[78(a)] Similarly, Chaudhary et al. [23] and Pistor et al. [79(b)] found that, overall, nanocomposite structures seemed independent of processing equipment. In addition, all nanocomposites appeared to be insensitive to different processing temperatures and shear forces.

3.4.2 EFFECT OF SILICATE TYPE ON DISPERSION OF EVA NANOCOMPOSITES

The effect of the silicate type on the dispersion of EVA composites was investigated by Zanetti et al. [73(b)], Riva et al. [74(a)], and Peeterbroeck et al. [77(b)] EVA matrices were melt blended with hectorite, fluorohectorite, magadiite [80], and MMT, all organically modified. The dispersion of hectorite-type clays was found to be better than that of MMT clays. This may be attributed to the large interfacial area available for the interaction of polymer chains with hectorite-like minerals. The interaction of silicates with a high surface area with polymer chains yields excellent dispersion.[23] Hectorite has an aspect ratio of approximately 5,000, whereas that of MMT is <1,000. Magadiite did not yield nanocomposites, only microcomposites.

3.4.3 EFFECT OF ORGANIC MODIFICATION OF SILICATES ON DISPERSION OF EVA NANOCOMPOSITES

Nanocomposite creation was found to depend on the type of silicate modification.[73] Suitable modification should render the silicate compatible with the polymer matrix. Silicates modified with ammonium cations bearing carboxylic acid moiety yielded conventional microcomposites, independent of the co-vinyl acetate content of EVA matrices. By contrast, organosilicates with nonfunctionalized alkyl ammonium tails (one and two) displayed affinity toward EVA chains. They yielded nanocomposites whatever the VA content (12, 19, and 27 wt %) of the matrix. Exfoliated silicate sheets were observed, together with stacks of intercalated and unmodified silicates. The nanocomposite structures were considered to be intercalated/exfoliated.[73] Higher numbers of stacks were observed in EVA12 nanocomposites. These authors speculated that such a small extent of exfoliation was due to the low polarity of the EVA12 matrix.

Similarly, organosilicates with one long alkyl chain yielded nanocomposites with relatively high numbers of stacks. Riva et al. [74(a)] used MMT modified with $(CH_2CH_2OH)_2N^+CH_3$(tallow) and EVA19. The polarity of MMT was increased, further improving its affinity toward polar EVA matrices. An exfoliated structure was reported for the nanocomposite based on EVA19. In addition, stacks of unmodified silicates were observed.[74(a)] Zhang et al. [8(b)] increased the number of long alkyl chains used to modify MMT up to three. They anticipated that such an increase would decrease the organo-MMT polarity. Nevertheless, they found it necessary to yield wider organo-MMT basal spacing. The aim was to obtain a proper balance between the two as this would facilitate the migration and penetration of EVA chains into the silicate layers. The morphological features of the EVA–MMT nanocomposites were then examined. They were found to depend on the basal spacing of the organically modified MMT and the polarity of the EVA. Increasing both promoted better dispersions. The chains of EVA diffused more easily into the MMT layers. Two long alkyl chains were found sufficient to yield organosilicates with wider basal spacing.[8(b)] These authors recorded further marginal expansion of d_{001} for triple-tailed organoclay.

3.4.4 EFFECT OF VA CONTENT ON DISPERSION OF EVA NANOCOMPOSITES

The effect of changing the matrix VA content on the dispersion of the EVA composites was assessed by several authors. [4,6,23,73,74(b),76,79(a)] Increasing the EVA polarity lowers the thermodynamic energy barrier for polymer interaction with silicates, and therefore the polymer chains diffuse more easily into the silicate layers. Increased matrix amorphousness (with increasing VA units) further facilitates the stabilization of polymer chains within the silicate galleries.[23] These authors claimed that higher amorphous content prevents recrystallization of polymer chains during annealing. Therefore, the chains remain diffused within the silicate layers.[23]

Whatever the VA content, organosilicates with OH groups along the alkyl N substituents appeared to be well dispersed. They may be due to strong intermolecular interactions between the OH groups of the organic modifier and the acetate functions of the EVA matrix.[77(b)] Increasing the VA content improved the degree of organosilicate dispersion, independent of the type of silicate modification. When the number of long tails was the same, organosilicates with higher chain lengths dispersed better. This may be related to the interlayer spacing of the organosilicate. Densely packing modifier into the silicate did not aid dispersion. Chaudhary et al. [23] claimed that this reduces the number of EVA chains penetrating the interlayer spaces.

Zhang and Sundararaj [6] investigated the extent of dispersion of some double-tailed organosilicates in EVA matrices with five VA contents (6, 9, 12, 18, and 28 wt %). It was found that all EVA matrices further expanded organoclay interlayers. Increasing the VA content from 6 to 12 wt % expanded the silicate interlayers considerably. Above such VA contents, no further interlayer expansion was recorded. An intercalation-limiting effect of the polarity after a certain critical VA content was revealed.[6] This critical VA content was found to be approximately 15 wt %. The degree of intercalation of EVA into double-tailed organoclay increased only at VA contents up to about 15 wt %. Thereafter the expansion of basal spacing ceased. The interlayer expansion was attributed to increased diffusion of EVA. Polymer diffusion depends strongly on how well it flows. The latter is determined using the melt flow index (MFI).

3.4.5 EFFECT OF MFI ON DISPERSION OF EVA NANOCOMPOSITES

The propensity of EVA with a higher VA content to diffuse into the silicate interlayers has been established.[8(b)] Zhang and Sundararaj [6] examined the influence of MFI on the structure of nanocomposites. They used five EVA28 matrices with different MFIs (3, 6, 25, 43, and 150 g/10 min). The effect of the MFI on the intercalation-limiting effect of EVA polarity into double-tailed organoclay was investigated. Lowering the MFI from 150 to 25 did not cause any detectable change in the basal spacing. However, further decreasing the MFI to 6 did expand the silicate interlayer. Below this MFI, the silicate interlayer collapsed. It was then concluded that effective polymer diffusion requires a suitable conjugation between its mobility and its shear force.[6] The shear force should (i) create shear tensions during nanocomposite processing; (ii) aid the breaking up of organosilicate agglomerations; and (iii) disperse silicate platelets or a few tactoids throughout the matrix and keep the silicate platelets or tactoids apart. With regard to polymer mobility, it should be sufficient to promote the diffusion and penetration of polymer into the silicate layers before layer restacking. The existence of an intercalation-limiting effect of EVA into double-tailed organoclays was confirmed, although it appears to be dependent on the MFI of the matrix rather than on its VA content. Zhang and Sundararaj [6] recorded increasing interlayer distances with increasing

VA content from 3 to 15 wt %. No further expansion was recorded for organo-MMT intercalated by EVA22 with different MFIs (2 and 3 g/10 min). Zhang and Sundararaj [6] also recorded increased interlayer spacing when the MFI of EVA28 was lowered from 25 to 6. Thus, it is speculated that EVA resins conjugating good mobility and sufficient shear force will have an MFI in the range 3–25. Marini et al. [79(a)] agreed that matrix viscosity is the driving force for polymer chain mobility within clay lamellae. In addition, imposed shear tension is also responsible for causing lamellae slippage and clay dispersion. Adequate affinity between polymer matrix and organosilicate was thus confirmed as indispensable.

3.5 EFFECT OF EVA NANOCOMPOSITE STRUCTURE ON ITS PHYSICAL PROPERTIES AND PERFORMANCE

3.5.1 INFLUENCE OF COMPOSITE STRUCTURE ON ITS MECHANICAL PROPERTIES

The effect of VA content on the mechanical properties of EVA/Mg LDH nanocomposites has been studied by various groups.[8(b),83] As expected, various nanocomposites exhibit a much higher storage modulus than pure EVA grades, especially at low temperatures, given the reinforcing effect of nanofillers on the matrix. In addition, the presence of the fillers enables the matrix to sustain high modulus values at high temperatures. In addition, various nanocomposites show a reduction in tan δ peak height compared with the heights of the respective neat elastomers. This is due to the restriction in polymer chain movements imposed by the filler–polymer interactions. The enhancements in dynamic mechanical properties indicate that the more elastomeric (VA content) the matrix is, the more easily the nanofillers are dispersed due to the higher free volume.

Typically, silicate particles have higher tensile moduli than polymer matrices. With increasing concentration of nanofiller, there is an increase in the Young's modulus (stiffness) of the nanocomposites. Alexandre and DuBois [84], using 5 wt % MMT, prepared nanocomposites with double the Young's modulus of pure EVA27. It was found that EVA19 and EVA12 increased the Young's modulus by 50 percent. The variation in the modulus of nanocomposites was explained on the basis of their different structures. The dispersion of individual clay platelets responsible for the large increase in modulus was higher in the EVA27 nanocomposite.[84] Apart from polymer polarity, silicate modification with a surfactant having nonfunctionalized chains compatible with polymer matrices was similarly critical. The ductility of the EVA27 nanocomposite decreased only slightly compared with that of the pure polymer. This was in spite of a large increase in nanocomposite stiffness.

Zhang and Sundararaj [6] recorded ever-increasing Young's moduli of EVA nanocomposites with increasing concentration of nanofiller. In parallel, they proposed the existence of a "platelet saturation effect." Such an effect reduces the extent of platelet dispersion in the polymer matrix. The saturation effect is explained as follows: Layered silicates have a large aspect ratio, exceeding 300; interaction between them is quite strong because of the large packing area; exfoliation and dispersion of silicate layers depend mainly on two factors: EVA–silicate interaction (ε_{es}) and silicate–silicate interaction (ε_{ss}); when $\varepsilon_{es} > \varepsilon_{ss}$, exfoliation of silicate layers is possible; conversely, when $\varepsilon_{es} < \varepsilon_{ss}$, exfoliation is impossible; an increase in clay content leads to a larger ε_{ss}; this is due to a shorter distance between the silicate aggregates.[6] The effect of the interplay between EVA polarity (amorphicity) and silicate concentration (wt %) on the Young's modulus has been evaluated. It has been accepted that platelet "randomization" characterizes exfoliated nanocomposite structures. Typically, the effective dispersion of nanosilicates suppresses the ability of the matrix to absorb energy at lower VA content in the EVA matrix.[23] Nanosilicates increase spatial hindrance for polymeric chain movement. They impart rigidity to the polymer matrix, creating a "rigid" amorphous phase. Platelet–polymer and platelet–platelet interactions tend to create a flexible silicate network structure in the matrix. Owing to polymer entanglement, such a network increases the initial resistance of polymeric chains moving under stress. The initial deformation energy is then absorbed by the silicate network. Simultaneously, the flexible network increases the modulus of the nanocomposite. With increasing VA content, the flexibility of the silicate network increases. Consequently, the resistance of the polymeric chains to movement is lowered. Thus, a "mobile" amorphous phase develops, and the network's ability to absorb deformation energy decreases. This occurs in spite of platelet–polymer and platelet–platelet interactions in the flexible silicate network. Stress is then partially transferred to the polymer chains, allowing them to absorb higher deformation energy. Hence, the modulus appears to be dominated by the extent of matrix crystallinity/amorphousness rather than by the silicate network.

Rigidity may also be imparted without the formation of a silicate network structure. There will be good interaction between silicate platelets or clusters of tactoids with the matrix where they are dispersed and suitably oriented. However, tensile strength is likely to be reduced.[6] Favorable interactions at the polymer/silicate interface are critical for efficient stress transfer. Tensile strength does not increase when polymer–clay interactions are sufficiently developed. The strength of the nanocomposite reduces with increasing flexibility of the silicate network structure. Increasing matrix polarity tends to maximize the extent of diffusion of EVA into silicate layers. A higher specific surface area becomes available for polymer–silicate interactions.[23]

3.5.2 INFLUENCE OF COMPOSITE STRUCTURE ON ITS STEADY SHEAR RHEOLOGICAL PROPERTIES

The degree of dispersion of silicates in a polymer matrix affects the rheological behavior of nanocomposites. Measurement of complex viscosity by oscillatory testing is useful to estimate the degree of exfoliation of composites. The viscosity of highly dispersed nanocomposites, with an exfoliated structure, increases considerably when the shear rate is changed. Conversely, the viscosity of poorly dispersed nanocomposites increases only moderately with the shear rate. At a low shear rate, exfoliated nanocomposites have the propensity to display solid-like behavior. This has been attributed to the formation of a network structure by dispersed silicate layers.[77(a)] Polymer chains are entrapped within the network. Because these are unable to flow, the viscosity rises. High zero-shear viscosities indicate that the network of dispersed layers remains unaffected by the imposed flow. Interactions between silicate layers and polymer chains are more pronounced in exfoliated systems than in fully intercalated systems. At the same silicate concentration, the elastic modulus is higher for exfoliated structures than for intercalated structures.[74(a)] Hence, solid-like behavior occurs at higher silicate loading in the latter systems. This leads to a slower relaxation of polymer chains.[77(a)]

High shear rates break down the silicate network and orient the platelets in the direction of flow. For this reason, nanocomposites exhibit shear-thinning behavior. The slope of curves, the so-called shear-thinning exponent, is used to estimate the extent of nanocomposite exfoliation. It has been accepted that higher absolute values of the exponent indicate higher rates of exfoliation. [77(a),78(b)] However, Marini et al. [79(a)] suggested that a significant increase in viscosity in the low shear region indicates strong matrix–organo-silicate interactions rather than exfoliation. Both well-dispersed intercalated and/or exfoliated silicates can lead to a huge increase in zero-shear viscosity. [79(a)]

La Mantia and Tzankova Dintcheva [78(a)] stated that the intensity of ma-trix–organosilicate interactions increases with silicate interlayer spacing. When the basal spacing increases, the surface area available for contact with the poly-meric chains also increases. Moreover, due to the larger interplatelet distances, the volume concentration of the silicate increases.[81,78(a)] High interactions between the organosilicate and the polymer chains are critical for nanocompos-ite creation. However, they are not sufficient on their own to guarantee effective clay dispersion and exfoliation.[6,79(a)] Strong matrix–organosilicate interac-tions are indicated by a significant increase in zero-shear viscosity, rather than simply high zero-shear viscosity. Marini et al. [79(a)] recorded large rheological differences between EVA nanocomposites, depending on the matrix viscosity. High-viscosity EVA12 (MFI = 0.3 g/10 min) and EVA19 (MFI = 2.1 g/10 min) consisted of fairly well-dispersed compact tactoids and had higher zero-shear viscosity than their respective EVA matrices. However, such viscosities were of the same order of magnitude or were only one order of magnitude different. Further, pure matrices also displayed pseudoplastic behavior. Absolute values of the "shear-thinning exponent" calculated for EVA12 nanocomposite and its ma-trix were high. Similarly, nanocomposites produced with low-viscosity EVA18 (MFI = 150 g/10 min) and EVA28 (MFI = 25 g/10 min) exhibited higher zero-shear viscosity than their respective EVA matrices. However, such viscosities were more than one order of magnitude different. Moreover, pure matrices ex-hibited Newtonian behavior. It was then concluded that organosilicate dispersion was dependent on EVA matrix polarity and viscosity.[79(a)] On its own, a high "shear-thinning exponent" does not guarantee a higher rate of exfoliation. Like-wise, a high zero-shear viscosity of the nanocomposites does not, on its own, guarantee strong matrix–organosilicate interactions.

3.5.3 INFLUENCE OF COMPOSITE STRUCTURE ON ITS FIRE PROPERTIES

The effect of several parameters (nature of clay and clay loading) on the fire retardancy of the nanocomposite has been investigated. It has been observed that the nature of the cations, which compensate for the negative charge of the silicate layers, affects the fire performance, even though the fire properties were improved for both the MMT-type fillers investigated. The clay loading also affects the fire properties.[75] The Stanton Red croft Cone Calorimeter was used to carry out measurements. The conventional data, namely time to ignition (TTI, s), heat release rate (HRR, kW/m^2), peak of heat release (PHRR, kW/m^2), that is, maximum of HRR, total heat release (THR, MJ/m^2), and WL (kg) were supplied by Polymer Laboratories software.

Huang et al. [52,56] explained the synergistic flame-retardant effects between sepiolite and magnesium hydroxide in EVA matrices. In light of the positive results from the loss on ignition and UL-94 tests, not only did the cone calorimeter test data indicate a reduction in the HRR and MLR, but also a prolonged TTI and a depressed smoke release were observed during combustion. Simultaneously, the tensile strength and Young's modulus of the system were also improved by the further addition of sepiolite due to the hydrogen bonds between silanol attached to sepiolite molecules and the ester groups of EVA.

Cárdenas et al. [85] studied the mechanical and fire-retardant properties of EVA/clay/sepiolite nanocomposites. Their results suggest that the synergistic effect is greater for bentonite with silica and with sepiolite than for bentonite with ATH. This is an expected effect in the case of bentonite with sepiolite taking into account that both inorganic fillers are phyllosilicates and have analogous chemical composition. However, the differences between the PHRR may also be influenced by the specific combustion mechanism of the different inorganic fillers used (silica, sepiolite). It is worth noting that EVA–sepiolite showed the lower THR among the other composites, confirming the synergistic effect between bentonite and sepiolite explained above. In the EVA–sepiolite composite, a uniform and rigid layer was formed, hence the contribution of sepiolite in the formation of a more rigid layer of char was very clear, possibly due to the fibril structure of this type of clay. Consequently, it is possible to conclude that a nanostructure enables better fire performance to be achieved than a microstructure. In fact, the presumed "diffusion effect," which leads to such improvements, occurs in a nanostructure but not in a macrostructure.

3.6 CONCLUSIONS

Detailed accounts of the different types of NSMs to enhance the novel properties of pristine EVA have been covered in this review. The review has been systematically structured to give a clear and detailed insight into the materials. In the Introduction, we reviewed recent papers on the subject and classified the NSMs into three categories according to their dimensions: 0DNSM (POSS), 1DNSM (CNT, sepiolite), and 2DNSM (clay) with EVA. In the next section, we presented a detailed discussion on the effects of POSS, various types of silicate structure, and various organic modifiers on EVA nanocomposites. In the third section, we discussed recent approaches to NSMs such as CNT, sepiolite, clay, and POSS. As NSMs play a vital role in EVA nanocomposites, we also elucidated the influence of composite structures on their thermal, mechanical, and fire-retardant properties.

With great progress being made in the preparation of EVA nanocomposites, there are fascinating new opportunities for materials scientists. Although considerable attention is being paid to particular aspects of nanostructures (for instance 0DNSM, 1DNSM, and 2DNSM), future progress will hinge on a better understanding of EVA nanocomposites, their composition, size, and morphology, which affect the activity of 0DNSM, 1DNSM, and 2DNSM. In addition, as greater knowledge is acquired about the physical and chemical properties of 0DNSM, 1DNSM, and 2DNSM, there will be more opportunities to exploit individual characteristics in thermal, electrical, mechanical, and fire-retardant-based applications. Moreover, the development of 0DNSM, 1DNSM, and 2DNSM will help to improve our old technologies, and further research will provide more impressive results that will benefit various industries and society. Finally, it is important to note that new types of cubic silica (POSS) nanoparticles have recently been reported and their ability to form nanocomposites with enhanced properties has been proposed.

ACKNOWLEDGMENTS

Financial support for this research from the Institutional Research Development Programme (IRDP), the South Africa/Mozambique Collaborative Programme of the National Research Foundation (NRF), and the Mozambican Research Foundation (FNI) is gratefully acknowledged. We are also grateful for the Vice-Chancellor's Postdoctoral fellowship, University of Pretoria, South Africa. The authors also acknowledge technical support from the Centre of Engineering Sciences at the Martin Luther University of Halle-Wittenberg.

KEYWORDS

- Clay
- CNT
- EVA
- POSS
- Sepiolite

REFERENCES

1. Hull, T. R.; Price, D.; Liu, Y.; Wills, C. L.; and Brady, J.; An investigation into the decomposition and burning behavior of ethylene-vinyl acetate copolymer nanocomposite materials. *Polym. Degrad. Stab.* **2003,** *82(2),* 365–371.

2. Gnanasekaran, D.; Madhavan, K.; Tsibouklis, J.; and Reddy, B. S. R.; Ring opening metathesis polymerization of polyoctahedral oligomeric silsesquioxanes (POSS) incorporated oxanorbornene-5,6-dicarboximide: synthesis, characterization, and surface morphology of copolymers. *Aust. J. Chem.* **2011,** *64(3),* 309–315.

3. George, J. J.; and Bhowmick, A. K.; Influence of matrix polarity on the properties of ethylene-vinyl acetate-carbon nanofiller nanocomposites. *Nanoscale. Res. Lett.* **2009,** *4(7),* 655–664.

4. Cui, L.; Ma, X.; and Paul, D. R.; Morphology and properties of nanocomposites formed from ethylene-vinyl acetate copolymers and organoclays. *Polym.* **2007,** *48(21),* 6325–6339.

5. Henderson Alex, M.; Ethylene-vinyl acetate (EVA) copolymers: a general review. *IEEE. Electr. Insul. Mag.* **1993,** *9(1),* 30–38.

6. Zhang, F.; and Sundararaj, U.; Nanocomposites of ethylene-vinyl acetate copolymer (EVA) and organoclay prepared by twin-screw melt extrusion. *Polym. Compos.* **2004,** *25(5),* 535–542.

7. Chaudhary, D. S.; Prasad, R.; Gupta, R. K.; and Bhattacharya, S. N., Morphological influence on mechanical characterization of ethylene-vinyl acetate copolymer-clay nanocomposites. *Polym. Eng. Sci.* **2005,** *45(7),* 889–897.

8. (a) Chiu, C.-W.; Huang, T.-K.; Wang, Y.-C.; Alamani, B. G.; and Lin, J.-J.; Intercalation strategies in clay/polymer hybrids. *Prog. Polym. Sci. 25(5),* **2013.**
   (b) Zhang, C.; Tjiu, W. W.; Liu, T.; Lui, W. Y.; Phang, I. Y.; and Zhang, W. D.; Dramatically enhanced mechanical performance of nylon-6 magnetic composites with nanostructured hybrid one-dimensional carbon nanotube-two-dimensional clay nanoplatelet heterostructures. *J. Phys. Chem. B.* **2011,** *115(13),* 3392–9.

9. (a) Kamarudin, S. K.; Achmad, F.; and Daud, W. R. W.; Overview on the application of direct methanol fuel cell (DMFC) for portable electronic devices. *Int. J. Hydrogen. Energy.* **2009,** *34(16),* 6902–6916.

(b) Chen, H.; Cong, T. N.; Yang, W.; Tan, C.; Li, Y.; and Ding, Y.; Progress in electrical energy storage system: a critical review. *Prog. Nat. Sci.* **2009**, *19(3)*, 291–312.

10. Ferreira-Aparicio, P.; Folgado, M. A.; and Daza, L.; High surface area graphite as alternative support for proton exchange membrane fuel cell catalysts. *J. Power. Sources.* **2009**, *192(1)*, 57–62.

11. Jin, Y. H.; Lee, S. H.; Shim, H. W.; Ko, K. H.; and Kim, D. W.; Tailoring high-surface-area nanocrystalline TiO2 polymorphs for high-power Li ion battery electrodes. *Electrochim. Acta.* **2010**, *55(24)*, 7315–7321.

12. (a) Dong, Z.; Kennedy, S. J.; and Wu, Y.; Electrospinning materials for energy-related applications and devices. *J. Power. Sources.* **2011**, *196(11)*, 4886–4904

    (b) Armand, M.; and Tarascon, J. M.; Building better batteries. *Nat.* **2008**, *451(7179)*, 652–657.

    (c) Simon, P.; and Gogotsi, Y.; Materials for electrochemical capacitors. *Nat. Mater.* **2008**, *7(11)*, 845–854.

    (d) Hu, C. C.; Chang, K. H.; Lin, M. C.; and Wu, Y. T.; Design and tailoring of the nanotubular arrayed architecture of hydrous RuO2 for next generation supercapacitors. *Nano Lett.* **2006**, *6(12)*, 2690–2695.

13. Shen, Q.; Jiang, L.; Zhang, H.; Min, Q.; Hou, W.; and Zhu, J. J.; Three-dimensional dendritic Pt nanostructures: sonoelectrochemical synthesis and electrochemical applications. *J. Phys. Chem. C.* **2008**, *112(42)*, 16385–16392.

14. (a) Teng, X.; Liang, X.; Maksimuk, S.; and Yang, H.; Synthesis of porous platinum nanoparticles. *Small.* **2006**, *2(2)*, 249–253.

    (b) Lee, H.; Habas, S. E.; Kweskin, S.; Butcher, D.; Somorjai, G. A.; and Yang, P.; Morphological control of catalytically active platinum nanocrystals. *Angew. Chemie.* **2006**, *45(46)*, 7824–8.

15. Brick, C. M.; Ouchi, Y.; Chujo, Y.; and Laine, R. M.; Robust polyaromatic octasilsesquioxanes from polybromophenylsilsesquioxanes, Br xOPS, via suzuki coupling. *Macromol.* **2005**, *38(11)*, 4661–4665.

16. Schwab, J. J.; and Lichtenhan, J. D.; Polyhedral oligomeric silsesquioxane (POSS)-based polymers. *Appl. Organomet. Chem.* **1998**, *12(10–11)*, 707–713.

17. Li, G.; Wang, L.; Ni, H.; and Pittman Jr, C. U.; Polyhedral oligomeric silsesquioxane (POSS) polymers and copolymers: a review. *J. Inorg Organomet. Polym.* **2001**, *11(3)*, 123–154.

18. Kudo, T.; Machida, K.; and Gordon, M. S.; Exploring the mechanism for the synthesis of silsesquioxanes. 4. The synthesis of T 8. *J. Phys. Chem. A.* **2005**, *109(24)*, 5424–5429.

19. Kuchibhatla, S. V. N. T.; Karakoti, A. S.; Bera, D.; and Seal, S.; One dimensional nanostructured materials. *Prog. Mater. Sci.* **2007**, *52(5)*, 699–913.

20. Li, Z.-M.; Li, S.-N.; Xu, X.-B.; and Lu, A.; Carbon nanotubes can enhance phase dispersion in polymer blends. *Polym. Plast. Technol. Eng.* **2007**, *46(2)*, 129–134.

21. Spitalsky, Z.; Tasis, D.; Papagelis, K.; and Galiotis, C.; Carbon nanotube-polymer composites: chemistry, processing, mechanical and electrical properties. *Prog. Polym. Sci. (Oxford).* **2010**, *35(3)*, 357–401.

22. Gorrasi, G.; Bredeau, S.; Candia, C. D.; Patimo, G.; Pasquale, S. D.; and Dubois, P.; Carbon nanotube-filled ethylene/vinylacetate copolymers: from *in situ* catalyzed

polymerization to high-performance electro-conductive nanocomposites. *Polym. Adv. Technol.* **2012**, *23(11),* 1435–1440.

23. Chaudhary, D. S.; Prasad, R.; Gupta, R. K.; and Bhattacharya, S. N.; Clay intercalation and influence on crystallinity of EVA-based clay nanocomposites. *Thermochim. Acta.* **2005**, *433(1–2),* 187–195.

24. Baughman, R. H.; Zakhidov, A. A.; de Heer, W. A., Carbon nanotubes—the route toward applications. *Sci.* **2002**, *297(5582),* 787–92.

25. Zheng, Y.; and Zheng, Y.; Study on sepiolite-reinforced polymeric nanocomposites. *J. Appl. Polym. Sci.* **2006**, *99(5),* 2163–2166.

26. Chen, H.; Zheng, M.; Sun, H.; and Jia, Q.; Characterization and properties of sepiolite/ polyurethane nanocomposites. *Mater. Sci. Eng.: A.* **2007**, *445–446,* 725–730.

27. (a) Tartaglione, G.; Tabuani, D.; Camino, G.; Moisio, M.; PP and PBT composites filled with sepiolite: morphology and thermal behaviour. *Compos. Sci. Technol.* **2008**, *68(2),* 451–460.

(b) Alkan, M.; and Benlikaya, R.; Poly(vinyl alcohol) nanocomposites with sepiolite and heat-treated sepiolites. *J. Appl. Polym. Sci.* **2009**, *112(6),* 3764–3774.

28. Huang, N. H.; Synergistic flame retardant effects between sepiolite and magnesium hydroxide in ethylene-vinyl acetate (EVA) matrix. *Express. Polym. Lett.* **2010**, *4(4),* 227–233.

29. Tiwari, J. N.; Tiwari, R. N.; and Kim, K. S.; Zero-dimensional, one-dimensional, two-dimensional and three-dimensional nanostructured materials for advanced electrochemical energy devices. *Prog. Mater. Sci.* **2012**, *57(4),* 724–803.

30. Jun, Y.; Seo, J.; Oh, S.; and Cheon, J.; Recent advances in the shape control of inorganic nano-building blocks. *Coord. Chem. Rev.* **2005**, *249(17–18),* 1766–1775.

31. Kim, K. S.; et al. Large-scale pattern growth of graphene films for stretchable transparent electrodes. *Nat.* **2009**, *457(7230),* 706–10.

32. Bae, S.; et al. Roll-to-roll production of 30-inch graphene films for transparent electrodes. *Nat. Nanotechnol.* **2010**, *5(8),* 574–578.

33. (a) Kurecic, M.; and Sfiligoj, M.; Polymer nanocomposite hydrogels for water purification. **2012**, *23(11),* 9–24.

(b) Chen, B.; et al. Critical appraisal of polymer-clay nanocomposites. *Chem. Soc. Rev.* **2008**, *37(3),* 568–594.

34. Chrissafis, K.; and Bikiaris, D.; Can nanoparticles really enhance thermal stability of polymers? part I: an overview on thermal decomposition of addition polymers. *Thermochim. Acta.* **2011**, *523(1–2),* 1–24.

35. Gnanasekaran, D.; and Reddy, B. S. R.; Synthesis and characterization of nanocomposites based on copolymers of POSS-ONDI macromonomer and TFONDI: effect of POSS on thermal, microstructure and morphological properties. *Adv. Mater. Res.* **2010**, *123–125,* 775–778.

36. Balazs, A. C.; Emrick, T.; and Russell, T. P.; Nanoparticle polymer composites: where two small worlds meet. *Sci.* **2006**, *314(5802),* 1107–1110.

37. (a) Krishnamoorti, R.; and Vaia, R. A.; Polymer nanocomposites. *J. Polym. Sci. Part B: Polym. Phys.* **2007**, *45(24),* 3252–3256

(b) Schaefer, D. W.; and Justice, R. S.; How nano are nanocomposites? *Macromol.* **2007**, *40(24),* 8501–8517.

38. Gnanasekaran, D.; Ajit Walter, P.; Asha Parveen, A.; and Reddy, B. S. R.; Polyhedral oligomeric silsesquioxane-based fluoroimide-containing poly(urethane-imide) hybrid membranes: Synthesis, characterization and gas-transport properties. *Sep. Purif. Technol.* **2013,** *111,* 108–118.

39. Yang, Y.; and Heeger, A. J.; A new architecture for polymer transistors. *Nat.* **1994,** *372(6504),* 344–346.

40. Gatos, K. G.; Martínez Alcázar, J. G.; Psarras, G. C.; Thomann, R.; and Karger-Kocsis, J., Polyurethane latex/water dispersible boehmite alumina nanocomposites: thermal, mechanical and dielectrical properties. *Compos. Sci. Technol.* **2007,** *67(2),* 157–167.

41. Fox, D. M.; et al. The pillaring effect of the 1,2-dimethyl-3(benzyl ethyl iso-butyl POSS) imidazolium cation in polymer/montmorillonite nanocomposites. *Polym.* **2011,** *52(23),* 5335–5343.

42. Vaia, R. A.; Jandt, K. D.; Kramer, E. J.; and Giannelis, E. P.; Kinetics of polymer melt intercalation. *Macromol.* **1995,** *28(24),* 8080–8085.

43. Fox, D. M.; et al. Use of a polyhedral oligomeric silsesquioxane (POSS)-imidazolium cation as an organic modifier for montmorillonite. *Langmuir.* **2007,** *23(14),* 7707–7714.

44. Zhao, F.; Bao, X.; McLauchlin, A. R.; Gu, J.; Wan, C.; and Kandasubramanian, B.; Effect of POSS on morphology and mechanical properties of polyamide 12/montmorillonite nanocomposites. *Appl. Clay. Sci.* **2010,** *47(3–4),* 249–256.

45. Tjong, S. C.; Structural and mechanical properties of polymer nanocomposites. *Mater. Sci. Eng. R.* **2006,** *53(3–4),* 73–197.

46. Ajayan, P. M.; Stephan, O.; Colliex, C.; and Trauth, D.; Aligned carbon nanotube arrays formed by cutting a polymer resin-nanotube composites. *Sci.* **1994,** *265,* 1212–1214.

47. Park, D.-H.; Hwang, S.-J.; Oh, J.-M.; Yang, J.-H.; and Choy, J.-H.; Polymer–inorganic supramolecular nanohybrids for red, white, green, and blue applications. *Prog. Polym. Sci.* **2013,** *38(10–11),* 1442–1486.

48. García-López, D.; Fernández, J. F.; Merino, J. C.; and Pastor, J. M.; Influence of organic modifier characteristic on the mechanical properties of polyamide 6/organosepiolite nanocomposites. *Compos. Part B: Eng.* **2013,** *45(1),* 459–465.

49. (a) Keledi, G.; Hari, J.; and Pukanszky, B., Polymer nanocomposites: structure, interaction, and functionality. *Nanoscale.* **2012,** *4(6),* 1919-38
    (b) Defontaine, G.; Barichard, A.; Letaief, S.; Feng, C.; Matsuura, T.; Detellier, C., Nanoporous polymer—clay hybrid membranes for gas separation. *J. Colloid Interface Sci.* **2010,** *343(2),* 622–7.

50. Basurto, F. C.; García-López, D.; Villarreal-Bastardo, N.; Merino, J. C.; and Pastor, J. M.; Nanocomposites of ABS and sepiolite: study of different clay modification processes. *Compos. Part B: Eng.* **2012,** *43(5),* 2222–2229.

51. Sinha Ray, S.; and Okamoto, M.; Polymer/layered silicate nanocomposites: a review from preparation to processing. *Prog. Polym. Sci. (Oxford).* **2003,** *28(11),* 1539–1641.

52. Scaffaro, R.; Botta, L.; Ceraulo, M.; and La Mantia, F. P.; Effect of kind and content of organo-modified clay on properties of PET nanocomposites. *J. Appl. Polym. Sci.* **2011,** *122(1),* 384–392.

53. Scaffaro, R.; Maio, A.; Agnello, S.; and Glisenti, A.; Plasma functionalization of multiwalled carbon nanotubes and their use in the preparation of nylon 6-based nanohybrids. *Plasma. Process. Polym.* **2012,** *9(5),* 503–512.

54. Pramanik, M.; Srivastava, S. K.; Samantaray, B. K.; and Bhowmick, A. K.; Synthesis and characterization of organosoluble, thermoplastic elastomer/clay nanocomposites. *J. Poly. Sci. Part B: Polym. Phys.* **2002,** *40(18),* 2065–2072.

55. Srivastava, S. K.; Pramanik, M.; and Acharya, H.; Ethylene/vinyl acetate copolymer/clay nanocomposites. *J. Poly. Sci. Part B: Polym. Phys.* **2006,** *44(3),* 471–480.

56. Huang, N. H.; Chen, Z. J.; Yi, C. H.; and Wang, J. Q.; Synergistic flame retardant effects between sepiolite and magnesium hydroxide in ethylene-vinyl acetate (EVA) matrix. *Express. Polym. Lett.* **2010,** *4(4),* 227–233.

57. Lee, H. M.; Park, B. J.; Choi, H. J.; Gupta, R. K.; Bhattachary, S. N., Preparation and rheological characteristics of ethylene-vinyl acetate copolymer/organoclay nanocomposites. *J. Macromol. Sci. Part B: Phys.* **2007,** *46(2),* 261–273.

58. Giannelis, E. P.; and Vaia, R. A.; Lattice model of polymer melt intercalation in organically-modified layered silicates. *Macromol.* **1997,** *30,* 7990–7999.

59. Costache, M. C.; Jiang, D. D.; and Wilkie, C. A.; Thermal degradation of ethylene–vinyl acetate coplymer nanocomposites. *Polym.* **2005,** *46(18),* 6947–6958.

60. Pradhan, S.; Costa, F. R.; Wagenknecht, U.; Jehnichen, D.; Bhowmick, A. K.; and Heinrich, G.; Elastomer/LDH nanocomposites: synthesis and studies on nanoparticle dispersion, mechanical properties and interfacial adhesion. *Eur. Polym. J.* **2008,** *44(10),* 3122–3132.

61. Wang, G.-A.; Wang, C.-C.; and Chen, C.-Y.; The disorderly exfoliated LDHs/PMMA nanocomposites synthesized by *in situ* bulk polymerization: the effects of LDH-U on thermal and mechanical properties. *Polym. Degrad. Stab.* **2006,** *91(10),* 2443–2450.

62. Zubitur, M.; Gómez, M. A.; and Cortázar, M.; Structural characterization and thermal decomposition of layered double hydroxide/poly(p-dioxanone) nanocomposites. *Polym. Degrad. Stab.* **2009,** *94(5),* 804–809.

63. Bocchini, S.; Morlat-Therias, S.; Gardette, J. L.; and Camino, G.; Influence of nanodispersed hydrotalcite on polypropylene photooxidation. *Eur. Polym. J.* **2008,** *44(11),* 3473–3481.

64. Magagula, B.; Nhlapo, N.; and Focke, W. W.; Mn2Al-LDH- and Co2Al-LDH-stearate as photodegradants for LDPE film. *Polym. Degrad. Stab.* **2009,** *94(6),* 947–954.

65. Allen, N. S.; Edge, M.; Rodriguez, M.; Liauw, C. M.; and Fontan, E.; Aspects of the thermal oxidation of ethylene vinyl acetate copolymer. *Polym. Degrad. Stab.* **2000,** *68,* 363–371.

66. Fina, A.; Tabuani, D.; Frache, A.; and Camino, G.; Polypropylene-polyhedral oligomeric silsesquioxanes (POSS) nanocomposites. *Polym.* **2005,** *46(19),* 7855–7866.

67. Zheng, L.; Waddon, A. J.; Farris, R. J.; and Coughlin, E. B.; X-ray characterizations of polyethylene polyhedral oligomeric silsesquioxane copolymers. *Macromol.* **2002,** *35(6),* 2375–2379.

68. Tanaka, K.; Adachi, S.; and Chujo Y.; Structure–property relationship of octa-substituted POSS in thermal and mechanical reinforcements of conventional polymers. *J. Polym. Sci.: Part A: Polym. Chem.* **2009,** *47,* 5690–5697.

69. Kopesky, E. T.; Haddad, T. S.; Cohen, R. E.; and McKinley, G. H.; Thermomechanical properties of poly(methyl methacrylate)s containing tethered and untethered polyhedral oligomeric silsesquioxanes. *Macromol.* **2004**, *37(24)*, 8992–9004.

70. Scapini, P.; et al. Thermal and morphological properties of high-density polyethylene/ ethylene-vinyl acetate copolymer composites with polyhedral oligomeric silsesquioxane nanostructure. *Polym. Int.* **2010**, *59(2)*, 175–180.

71. George, J.; and Bhowmick, A. K.; Fabrication and properties of ethylene vinyl acetate-carbon nanofiber nanocomposites. *Nanoscale Res. Lett.* **2008**, *3(12)*, 508–15.

72. Beyer, G.; Flame retardant properties of EVA-nanocomposites and improvements by combination of nanofillers with aluminium trihydrate. *Fire Mater.* **2001**, *25(5)*, 193–197.

73. (a) Michae Alexandre, G. B.; Catherine henrist, rudi cloots, andre´ rulmont, robert jerome, and philippe dubois, "one-pot" preparation of polymer/clay nanocomposites starting from Na+montmorillonite. 1. Melt intercalation of ethylene-vinyl acetate copolymer. *Chem. Mater.* **2001**, *13*, 3830–3832.

     (b) Zanetti, M.; Camino, G.; Thomann, R.; Mülhaupt, R.; Synthesis and thermal behaviour of layered silicate-EVA nanocomposites. *Polym.* **2001**, *42(10)*, 4501–4507.

74. (a) Riva, A.; Zanetti, M.; Braglia, M.; Camino, G.; Falqui, L., Thermal degradation and rheological behaviour of EVA/montmorillonite nanocomposites. *Polym. Degrad. Stab.* **2002**, *77(2)*, 299–304.

     (b) Cser, F.; and Bhattacharya, S. N.; Study of the orientation and the degree of exfoliation of nanoparticles in poly(ethylene-vinyl acetate) nanocomposites. *J. Appl. Polym. Sci.* **2003**, *90(11)*, 3026–3031.

75. Duquesne, S.; Jama, C.; Le Bras, M.; Delobel, R.; Recourt, P.; and Gloaguen, J.M.; Elaboration of EVA–nanoclay systems—characterization,thermal behaviour and fire performance. *Compos. Sci. Technol.* **2003**, *63*, 1141–1148.

76. Pasanovic-Zujo, V.; Gupta, R. K.; and Bhattacharya, S. N., Effect of vinyl acetate content and silicate loading on EVA nanocomposites under shear and extensional flow. *Rheol. Acta.* **2004**, *43(2)*, 99–108.

77. (a) Gupta, R. K.; Pasanovic-Zujo, V.; and Bhattacharya, S. N., Shear and extensional rheology of EVA/layered silicate-nanocomposites. *J. Non-Newtonian. Fluid. Mech.* **2005**, *128(2–3)*, 116–125.

     (b) Peeterbroeck, S.; Alexandre, M.; Jérôme, R.; and Dubois, P.; Poly(ethylene-*co*-vinyl acetate)/clay nanocomposites: effect of clay nature and organic modifiers on morphology, mechanical and thermal properties. *Polym. Degrad. Stab.* **2005**, *90(2)*, 288–294.

78. (a) La Mantia, F. P.; and Tzankova Dintcheva, N.; Eva copolymer-based nanocomposites: rheological behavior under shear and isothermal and non-isothermal elongational flow. *Polym. Test.* **2006**, *25(5)*, 701–708.

     (b) Szép, A.; Szabó, A.; Tóth, N.; Anna, P.; and Marosi, G., Role of montmorillonite in flame retardancy of ethylene–vinyl acetate copolymer. *Polym. Degrad. Stab.* **2006**, *91(3)*, 593–599.

79. (a) Marini, J.; Branciforti, M. C.; and Lotti, C., Effect of matrix viscosity on the extent of exfoliation in EVA/organoclay nanocomposites. *Polym. Adv. Technol.* **2009**, *40(10)*, 4501–4507.

(b) Pistor, V.; Lizot, A.; Fiorio, R.; and Zattera, A. J.; Influence of physical interaction between organoclay and poly(ethylene-*co*-vinyl acetate) matrix and effect of clay content on rheological melt state. *Polym.* **2010,** *51(22),* 5165–5171.

80. Filippi, S.; Paci, M.; Polacco, G.; Dintcheva, N. T.; and Magagnini, P.; On the interlayer spacing collapse of cloisite® 30B organoclay. *Polym. Degrad. Stab.* **2011,** *96(5),* 823–832.

81. Joseph, S.; and Focke, W. W.; Poly(ethylene-vinyl-*co*-vinyl acetate)/clay nanocomposites: mechanical, morphology, and thermal behavior. *Polym. Compos.* **2011,** *32(2),* 252–258.

82. Pavlidou, S.; and Papaspyrides, C. D.; A review on polymer-layered silicate nanocomposites. *Prog. Polym. Sci. (Oxford).* **2008,** *33(12),* 1119–1198.

83. Soon Suh, S. H. R.; Jong Hyun Bae; and Young-Wook Chang; Effects of compatibilizer on the layered silicate/ethylene vinyl acetate nanocomposite. *J. Appl. Polym. Sci.* **2004,** *94,* 1057–1061.

84. Alexandre, M.; and Dubois, P.; Polymer-layered silicate nanocomposites: preparation, properties and uses of a new class of materials. *Mater. Sci. Eng. R.* **2000,** *28(1),* 1–63.

85. Cárdenas, M. Á.; Basurto, F. C.; García-López, D.; Merino, J. C.; and Pastor, J. M.; Mechanical and fire retardant properties of EVA/clay/ATH nanocomposites: effect of functionalization of organoclay nanofillers. *Polym. Bull.* **2013,** *70(8),* 2169–2179.

CHAPTER 4

PORE STRUCTURE OF POLY(2-HYDROXYETHYL METHACRYLATE) SCAFFOLDS

D. HORÁK and H. HLHDKOVÁ

CONTENTS

4.1 INTRODUCTION

Polymer supports have received much attention as a microenvironment for cell adhesion, proliferation, migration, and differentiation in tissue engineering and regenerative medicine. The three-dimensional scaffold structure provides support for high level of tissue organization and remodeling. Regeneration of different tissues, such as bone [1], cartilage [2], skin [3], nerves [4], or blood vessels [5], is investigated using such constructs. An ideal polymer scaffold should thus mimic the living tissue, that is, possess high water content, with the possibility to incorporate bioactive molecules allowing a better control of cell differentiation. At the same time, it requires a range of properties including biocompatibility and/or biodegradability, highly porous structure with communicating pores allowing high cell adhesion and tissue in-growth. The material should be sterilizable and also possess good mechanical strength. Both natural and synthetic hydrogels are being developed. The advantage of synthetic polymer matrices consists in their easy proccessability, tunable physical and chemical properties, susceptibility to modifications, and possibly controlled degradation.

Many techniques have been developed to fabricate highly porous constructs for tissue engineering. They include solvent casting [6], gas foaming [7] or/and salt leaching [8], freeze–thaw procedure [9,10], supercritical fluid technology [11] (disks exposed to CO_2 at high pressure), and electrospinning (for nanofiber matrices).[12] A wide range of polymers was suggested for scaffolds. In addition to natural materials, such as collagen, gelatine, dextran [13], chitosan [14], phosphorylcholine [15], alginic [16], and hyaluronic acids, it also includes synthetic polymers, for example, poly(vinyl alcohol) [17], poly(lactic acid) [1,18], polycaprolactone [19], poly(ethylene glycol) [20], polyacrylamide [21], polyphosphazenes [22], and polyurethane.[23]

Among the various kinds of materials being used in biomedical and pharmaceutical applications, hydrogels composed of hydrophilic polymers or copolymers find a unique place. They have a highly water-swollen rubbery three-dimensional structure, which is similar to natural tissue.[24,25] In this report, poly(2-hydroxyethyl methacrylate) (PHEMA) was selected as a suitable hydrogel intended for cell cultivation. The presence of hydroxy and carboxy groups makes this polymer compatible with water, whereas the hydrophobic methyl groups and backbone impart hydrolytic stability to the polymer and support the mechanical strength of the matrix.[26] PHEMA hydrogels are known for their resistance to high temperatures, acid and alkaline hydrolysis, and low reactivity with amines.[27] Previously, porous structure in PHEMA hydrogels was obtained by phase separation using a low molecular weight or polymeric porogen, or by the salt-leaching method. The material was used as a mouse embryonic stem cell support.[8,28,29] The aim of this report is to demonstrate conditions under which communicating pores are formed enabling high permeability of PHEMA scaffolds, which is crucial for future cell seeding.

4.2 EXPERIMENTAL

4.2.1 REAGENTS

2-Hydroxyethyl methacrylate (HEMA; Röhm GmbH, Germany) and ethylene dimethacrylate (EDMA; Ugilor S.A., France) were purified by distillation. 2,2'-Azobisisobutyronitrile (AIBN, Fluka) was crystallized from ethanol and used as initiator. Sodium chloride G.R. (Lach-Ner, s.r.o. Neratovice, Czech Republic) was classified; particle size 250–500 μm and ammonium sulfate needles (100 × 600 μm, Lachema, Neratovice) were utilized as porogens. Cyclohexanol (CyOH, Lachema, Neratovice) was distilled; dodecan-1-ol (DOH) and all other solvents and reagents were obtained from Aldrich and used without purification. Ammonolyzed PGMA microspheres (2 μm) were obtained by the previously described procedure.[30] Sulfonated polystyrene (PSt) microspheres (8 μm) Ostion LG KS 0803 were purchased from Spolek pro chemickou a hutní výrobu, Ústí n. L., Czech Republic. Polyaniline hydrochloride microspheres (PANI, 200–400 nm) were prepared according to literature.[31]

4.2.2 HYDOGEL PREPARATION

Crosslinked hydrogel constructs were prepared by the bulk radical polymerization of a reaction mixture containing monomer (HEMA), crosslinking agent (EDMA), initiator (AIBN), and NaCl or/and liquid diluent as a porogen (CyOH/DOH = 9/1 w/w). The compositions of polymerization mixtures are summarized in Table 1. The amount of crosslinker (2 wt %) and AIBN (1 wt %) dissolved in monomers was the same in all experiments, whereas the amount of porogen in the polymerization batch varied from 35.9 to 41.4 vol %. Optionally, needle-like $(NH_4)_2SO_4$ crystals (42.3 vol %) together with saturated $(NH_4)_2SO_4$ solution were used as a porogen instead of NaCl crystals, allowing the formation of hydrogels with communicating pores (Run 9, Table 1). For the sake of comparison, a copolymer was prepared with a mixture of solid (NaCl) and liquid low-molecular-weight porogen (diluent), amounting to 50 percent of the polymerization feed (Run 8). The thickness of the hydrogel was adjusted with a 3-mm-thick silicone rubber spacer between the teflon plates (10 × 10 cm), greased with a silicone oil and covered with cellophane. The reaction mixture was transferred onto a hollow plate and covered with a second plate, clamped and heated at 70°C for 8 hours. After polymerization, the hydrogels obtained with inorganic salt as a porogen were soaked in water and washed until the reaction of chloride or sulfate ions disappeared. The scaffolds prepared in the presence of a liquid diluent were washed with ethanol/water mixtures (98/2, 70/30, 40/60, 10/90 v/v) and water to remove the diluent, unreacted monomers, and initiator residues. The washing water was changed every day for 2 weeks.

TABLE 1   Preparation of PHEMA hydrogels, conditions, and properties[a]

Run	NaCl (vol. %)	Water regain (ml/g)		CX regain[d] (ml/g)	Cumulative pore volume[f] (ml/g)
1	41.4	0.84[d]	7.52[e]	0.13	0.35
2	40.8	0.88[d]	7.40[e]	0.23	0.47
3	40.0	1.04[d]	5.34[e]	0.56	1.03
4	39.1	0.81[d]	4.05[e]	0.33	1.24
5	37.9	0.78[d]	4.04[e]	0.21	1.60
6	37.0	0.79[d]	3.64[e]	0.34	1.83
7	35.9	0.75[d]	3.32[e]	0.32	1.70
8[b]	37.9	0.89[d]	4.30[e]	0.45	1.65
9	42.3[c]	0.84[d]	2.11[e]	0.08	0.08

[a]Crosslinked with 2 wt % EDMA, 1 wt % AIBN, NaCl in vol % relative to polymerization mixture (HEMA + EDMA + NaCl).
[b]One half of the HEMA/EDMA feed was replaced by CyOH/DOH = 9/1 w/w.
[c]$(NH_4)_2SO_4$.
[d]Centrifugation method.
[e]Suction method.
[f]Mercury porosimetry.

4.2.3 METHODS

4.2.3.1 MICROSCOPY

Low-vacuum scanning electron microscopy (LVSEM) was performed with a microscope Quanta 200 FEG (FEI, Czech Republic). Neat hydrated hydrogels were cut with a razor blade into ~5 mm cubes, flash-frozen in liquid nitrogen, and placed on the sample stage cooled to −10°C. Before microscopic observation, the top of a frozen sample was cut off using a sharp blade. During the observation, the conditions in the microscope (−10°C, 100 Pa) caused slow sublimation of ice from the sample, which made it possible to visualize its 3-D morphology. All samples were observed with a low-vacuum secondary electron detector, using the accelerating voltage 30 kV. Lyophilized PHEMA hydrogels Runs 3 and 8 filled with microspheres were also examined using LVSEM; however, the microspheres were washed out of the pores during freezing and, consequently, were scarcely observed on the micrographs.

High-vacuum SEM (HVSEM) was carried out with an electron microscope Vega TS 51355 (Tescan, Czech Republic). Permeability of the water-imbibed hydrogels (Runs 7 and 9) was investigated by the flow of water suspension of the polymer microspheres. Before observation, the wet hydrogel was placed on a wet filtration paper and a droplet of a suspension of 8 μm PSt microspheres in water was placed on the top. The sample was dried at ambient temperature and cut with a sharp blade in the direction of the microsphere flow. Samples showing the top, bottom, and cross-sections of flowed-through hydrogels were sputtered with a 8 nm layer of platinum using a vacuum sputter coater (Baltec SCD 050), fixed with a conductive paste to a brass support, and viewed in a scanning electron microscope in high vacuum (10^{-3} Pa), using the acceleration voltage 30 kV and a secondary-electrons detector. This technique made it possible to observe both microspheres and pores of the hydrogel.

4.2.3.2 SOLVENT REGAIN

The solvent (water or cyclohexane—CX) regain was determined in 1×2 cm sponge pieces of hydrogel kept for 1 week in deionized water, which was exchanged daily. Water regain was measured by two methods: (i) centrifugation [31] (WR_c) and (ii) suction (WR_s). In centrifugation method, solvent-swollen samples were placed into glass columns with fritted disc, centrifuged at 980 g for 10 min and immediately weighed (w_w—weight of hydrated sample), then vacuum dried at 80°C for 7 hours and again weighed (w_d—weight of dry sample). In the second method, excessive water was removed from the imbibed hydrogel by suction and the hydrogel weighted to determine w_w. Weight of dry sample w_d was determined as above. Water-regained WR_c or WR_s (ml/g) were calculated according to the equation:

$$WR_c \left(WR_s \right) = \frac{w_w - w_d}{w_d} \tag{1}$$

The results are average values of two measurements for each hydrogel. To measure cyclohexane regain (CXR) by centrifugation, equilibrium water-swollen hydrogels were successively washed with ethanol, acetone, and finally cyclohexane. Using the solvent-exchange, a thermodynamically good (swelling) solvent in the swollen gel was replaced by a thermodynamically poor solvent (nonsolvent). Porosity of the hydrogels (p) was calculated from the water and CXRs (Table 1) and PHEMA density ($\rho = 1.3$ g/ml) according to the following equation:

$$p = \frac{R \times 100}{R + \frac{1}{\rho}} \; (\%) \tag{2}$$

where $R = WR_c$, WR_s, or CXR (ml/g).

4.2.3.3   MERCURY POROSIMETRY

Pore structure of freeze-dried PHEMA scaffolds was characterized on a mercury porosimeter Pascal 140 and 440 (Thermo Finigan, Rodano, Italy). It works in two pressure intervals, 0–400 kPa and 1–400 MPa, allowing determination of meso- (2–50 nm), macro- (50–1,000 nm), and small superpores (1–116 μm). The pore volume and most frequent pore diameter were calculated under the assumption of a cylindrical pore model using the PASCAL program. It employed Washburn's equation describing capillary flow in porous materials.[33] The volumes of bottle and spherical pores were evaluated as the difference between the end values on the volume/pressure curve. Porosity was calculated according to Eq. (2), where cumulative pore volume (meso-, macro-, and small superpores) from mercury porosimetry was used for R.

4.3.   RESULTS AND DISCUSSION

4.3.1 MORPHOLOGY OF HYDROGELS

The prepared PHEMA constructs always had an opaque appearance indicating a permanent porous structure. Pores are generally divided into micro-, meso-, macro-, small, and large superpores. Morphology of water-swollen PHEMA hydrogels was investigated using LVSEM as shown in Figure 1. Large 200–500 μm superpores were developed as imprints of NaCl crystals, which were subsequently washed from the hydrogel; the interstitial space between them was filled with the polymer. During the observation, ice crystals filling soft polymer net were clearly visible in the center of the hydrogel Run 1 (Table 1) prepared at the highest content of NaCl (41.4 vol %) in the feed (Figure 1(a)). The internal surface area was too small to be determined. Figures 1(b) and (c) show hydrogels from Runs 3 and 5 (40 and 37.9 vol % NaCl), respectively, documenting their more compact structure accompanied by thicker walls between large superpores as compared with the hydrogel from Run 1 (Figure 1(a)). According to LVSEM, ~8 μm pores, the presence of which was confirmed by mercury porosimetry (volume about 1 ml/g), were observed in the walls between the large superpores. Longitudinal cracks in the material structure (Figure 1(b)) were obviously caused by sample handling and fast freezing in liquid nitrogen. Nevertheless, the LVSEM micrographs displayed only cross-sections of hydrogels, and it was not clear whether their pores are interconnected.

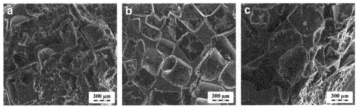

FIGURE 1   LVSEM micrographs showing frozen cross-section of PHEMA hydrogels prepared with (a) 41.4 vol %—Run 1, (b) 40 vol %—Run 3, and (c) 37.9 vol % NaCl (250–500 μm)—Run 5. PHEMA crosslinked with 2 wt % EDMA (relative to monomers).

Interconnection of pores is of vital importance for cell ingrowths in future applications to tissue regeneration. This feature was tested by the permeability of the whole hydrogels for different kinds of microspheres under two microscopic observations. First, cross-sections of the frozen hydrogels filled with microspheres were observed in LVSEM. Second, the water-swollen hydrogels were flowed through by a suspension of microspheres in water and their dried cross-sections were then viewed in HVSEM.

LVSEM showed water-swollen morphology of hydrogel constructs that preserved due to their freezing in liquid nitrogen. PHEMA Run 3 prepared with neat NaCl (Figure 2(a)) was compared with Run 8 obtained in the presence of NaCl together with a mixture of CyOH/DOH (Figure 2(d)). Addition of liquid porogens did not change the morphology; however, it increased the pore volume (from 0.21 to 0.45 ml CX/g, Table 1) and softness of the hydrogel. As a result, it had a tendency to disintegrate during the washing procedure. LVSEM of both PHEMA hydrogels filled with 2-μm ammonolyzed PGMA and 200–400 nm PANI microspheres is illustrated in Figures 2(b) and (e) and Figures 2(c) and (f), respectively. The micrographs showed undistorted morphology of the frozen hydrogels, but just a few microspheres and/or their agglomerates. This was attributed to the fact that most of them were washed out during preparation of the sample for LVSEM.

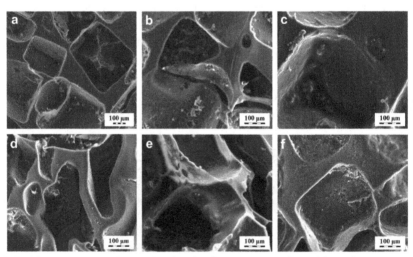

FIGURE 2   LVSEM micrographs showing frozen cross-section of PHEMA constructs: (a–c)—Run 3, (d–f)—Run 8; (a, d) neat and filled with (b, e) 2 μm ammonolyzed PGMA, and (c, f) 200–400 nm PANI microspheres.

Morphology of the PHEMA hydrogels flowed through by a suspension of microspheres was observed using HVSEM. Figure 3 shows HVSEM micrographs of cross-sections of the top and bottom part of the hydrogels from Runs 3, 5, and 8 flowed through by a suspension of 8-μm sulfonated PSt microspheres in water. Although the microspheres flowed through the hydrogel construct from Run 3, they did not penetrate the constructs from Runs 5 and 8 prepared in the presence of a rather low content of NaCl (37.9 vol %). At the same time, surface and inner structure of the hydrogels slightly differed. Figure 4 shows HVSEM of the longitudinal section of the PHEMA scaffold obtained with cubic NaCl crystals as a porogen (Run 7). Although Figure 4(a) shows the bulk, Figures 4(b) through (d) detailed sections. Again, small superpores with an average size about 13 μm were in the walls between the large superpores, forming small channels through which water flowed. To prove or exclude the interconnection of at least some pores, a suspension of 8 μm sulfonated polystyrene (PSt) microspheres in water was poured on the center of the top side of the gel. Although water flowed through the hydrogel bulk, the microspheres were retained on the surface of the hydrogel or penetrated only superficial layers due to the surface cracks (Figures 4(a) and (b)). This confirmed that the pores of PHEMA hydrogels obtained with a low content of NaCl porogen (35.9 vol %) did not communicate. By contrast, Figure 5 presents longitudinal section of the PHEMA hydrogel from Run 9 (both bulk and detailed) obtained with needle-like $(NH_4)_2SO_4$ crystals as a porogen. This porogen allowed formation of connected pores, which is explained by the needle-like structure of ammonium sulfate crystals that are linked to the gel structure. At the same time, the crystals grew to large structures due to the presence of saturated $(NH_4)_2SO_4$ solution in the feed. As a result, long interconnected large superpores—channels—were formed. This is documented in Figures 5(a) and (d) by the fact that suspension of 8 μm sulfonated PSt microspheres in water deposited in the center of the top side of the hydrogel flowed through. The captured microspheres are clearly visible in Figures 5(b) and (d). They accumulated at the places of pore narrowing; the majority, however, was found on the bottom part of the hydrogel. In this manner, the flow of water suspension of microspheres in the hydrogel was traced.

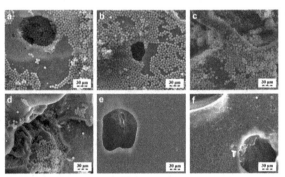

FIGURE 3   HVSEM micrographs of PHEMA hydrogels Run 3 (a, d), Run 5 (b, e), and Run 8 (c, f) showing top (a–c) and bottom (d–f) of the hydrogels after the flow of a suspension of 8 µm sulfonated PSt microspheres in water.

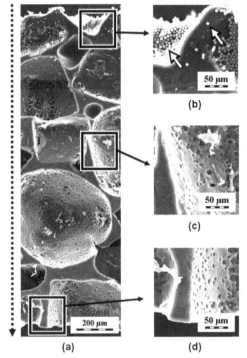

FIGURE 4   Selected HVSEM micrographs showing longitudinal section of PHEMA hydrogel 3 mm thick (Run 7) obtained with NaCl (250–500 µm) as a porogen after passing of a suspension of 8 µm sulfonated PSt microspheres in water (in the direction of the dotted line). (a) The whole cross-section through the hydrogel and selected details from, (b) top, (c) center, and (d) bottom. PSt microspheres are denoted with white arrows.

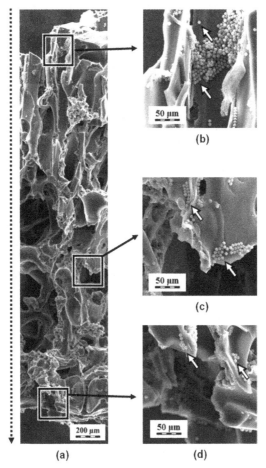

FIGURE 5 Selected HVSEM micrographs showing longitudinal section of PHEMA hydrogel 3 mm thick (Run 9) obtained with $(NH_4)_2SO_4$ (100×600 µm) as a porogen after passing of a suspension of 8 µm sulfonated PSt microspheres in water (in the direction of the dotted line). (a) The whole section through the construct and selected details from, (b) top, (c) center, and (d) bottom. PSt microspheres are denoted with white arrows.

Mechanical properties of the porous constructs were sensitive to the concentration of porogen in the feed. Hydrogels with lower contents of NaCl; and therefore, higher proportion of PHEMA had thicker walls between the pores and were more compact allowing increased swelling of polymer chains in water. Two PHEMA hydrogels with the highest contents of NaCl in the feed (41.4 vol %—Run 1 and 40.8 vol %—Run 2) possessing thin polymer walls between large superpores easily disintegrated as well as hydrogel prepared using $(NH_4)_2SO_4$ (42.3 vol %—Run 9).

4.3.2 CHARACTERIZATION OF POROSITY BY SOLVENT REGAIN

Dependencies of porosity of PHEMA hydrogels calculated from water or CXR and also from mercury porosimetry on NaCl content in the polymerization feed showed similar behavior (Figure 6). Porosities 81–91 and 49–57 percent for water regain were obtained by suction and centrifugation, respectively, 14–42 percent for CXR, and 31–70 percent for mercury porosimetry. The porosity determined by centrifugation of samples soaked with water and cyclohexane (solvents with different affinities to polar methacrylate chain) consists of two contributions: filling of the pores and swelling (solvation) of PHEMA chains. The uptake of cyclohexane, a thermodynamically poor solvent, which cannot swell the polymer, is a result of the former contribution only, reflecting thus the pore volume. The water regain from centrifugation was always higher than the CXR, demonstrating thus swelling of polymer chains with water (Table 1). Solvent regains were affected by the concentration of NaCl porogen in the polymerization feed. Porosities according to both water and CXRs by centrifugation slightly increased with increasing volume of NaCl porogen in the polymerization feed from 35.9 to 40 vol % and then decreased with a further NaCl increase up to 41.4 vol % (Figure 6). In the latter range of NaCl, the porosity evaluated using mercury porosimetry exhibited an analogous dependence. This decrease in solvent and mercury regains can be explained by thin polymer walls between large superpores inducing collapse of the porous structure. In the concentration range of NaCl in the feed 35.9–40 vol %, mercury porosimetry provided higher porosities than those obtained from regains by centrifugation at 980 g because it obviously did not retain solvents in large superpores. Retained water reflected thus only small superpores, closed pores, and solvation of the polymer in water similarly as observed earlier for macroporous PHEMA scaffolds.[34] Water regain was determined also by the suction method (Table 1), which showed values several times higher (3.3–7.5 ml/g) than by centrifugation due to filling all the pores in the polymer structure, including large superpores. As expected, porosity by the suction method increased with increasing volume of NaCl porogen in the polymerization feed (Figure 6).

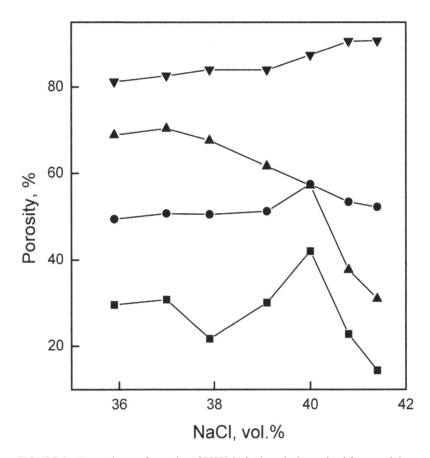

FIGURE 6 Dependence of porosity of PHEMA hydrogels determined from cyclohexane (■) and water regain measured by centrifugation (●) or suction (▼) and mercury porosimetry (▲) on the content of NaCl (250–500 μm) porogen in the polymerization feed.

The hydrogel from Run 8 formed in the presence of NaCl and CyOH/ DOH porogen showed higher solvent regains and mercury penetration than the comparable hydrogel from Run 5 obtained with the same content of neat NaCl (Table 1). This can be explained by the higher total amount of porogen in the former hydrogel. By contrast, the hydrogel from Run 9 prepared with needle-like $(NH_4)_2SO_4$ crystals as a porogen had the lowest solvent and mercury regains of all the samples. The exception was water regained by centrifugation, which was identical with that of sample Run 1 (Table 1) having a similar content of the NaCl porogen in the feed. This can imply that only large continuous superpores were present in this hydrogel and small superpores and macro- and mesopores were almost absent as evidenced by the low values of solvent and mercury regains.

4.3.3   CHARACTERIZATION OF POROSITY BY MERCURY POROSIMETRY

The advantage of mercury porosimetry is that it provides not only pore volumes, but also pore size distribution not available in other techniques. The method measures samples dried by lyophilization, which does not distort the pore structure. As mentioned earlier, porosities determined by mercury porosimetry were lower than those obtained from water regain by suction, which included large superpores, and higher than those from water and CXR detected by centrifugation. This was due to better filling of the compact xerogel structures obtained at lower contents of NaCl in the feed with mercury under a high pressure than with water or cyclohexane under atmospheric pressure. Figure 7 shows the dependence of most frequent mesopore size of PHEMA scaffolds and their pore volumes on the NaCl porogen content in the polymerization feed. Predominantly, 4–5 nm mesopores were detected with their volume increasing from 0.03 to 0.1 ml/g with increasing NaCl content in the polymerization feed. Macropores were absent and very low values of specific surface areas (<0.1 m^2/g) were found. The presence of CyOH/DOH porogen (Run 8) did not substantially affect the formation of meso- and macropores (volume 0.022 ml/g), because the amount of crosslinker in the polymerization feed was limited to only 2 wt %. The separation of the polymer from the porogen phase could not thus occur and porous structure was not formed. Both in hydrophobic styrene–divinylbenzene [35,36] and polar methacrylate copolymers [37] prepared in the presence of liquid porogens, phase separation and formation of macroporous structure occurred at crosslinker contents higher than 10 wt %.

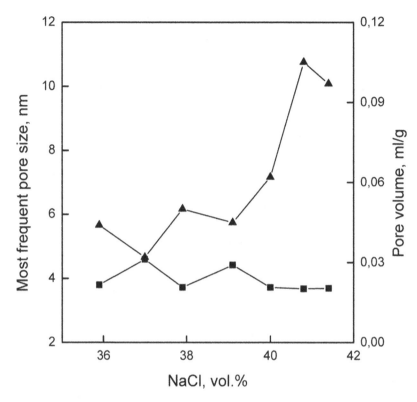

FIGURE 7 Dependence of pore volume (▲) and most frequent mesopore size (■) of PHEMA hydrogels on the content of NaCl (250–500 μm) porogen in the polymerization feed according to mercury porosimetry.

Figure 8 represents the dependence of most frequent small superpore size of PHEMA constructs and their pore volume on the content of NaCl porogen in the polymerization. Pore size increased up to 28–69 μm with increasing NaCl volume. This was maximum in the range 40–41.4 vol % NaCl probably due to the aggregation of NaCl crystals in the mixture at their high contents. All the investigated samples contained small superpores, the volume of which was about 20 times higher than that of mesopores. The volume of small superpores continuously decreased from 1.8 to 0.2 ml/g with rising NaCl amount in the feed. This could be explained by collapse of the pore structure and destruction of last two hydrogels with the highest content of NaCl porogen (Runs 1 and 2) under high pressure as mentioned earlier. The size of small superpores according to mercury porosimetry was by an order of magnitude smaller than the particle size of the used NaCl porogen (250–500 μm) because the method was able to distinguish the superpores only in the size range 1–116 μm (Figure 9). Large superpores (imprints of NaCl crystals) were detected using LVSEM. Figure 9 shows a typical cumulative pore volume and a derivative pore size distribution curve of PHEMA Run 4 with a decisive contribution of small superpores 25 μm in size.

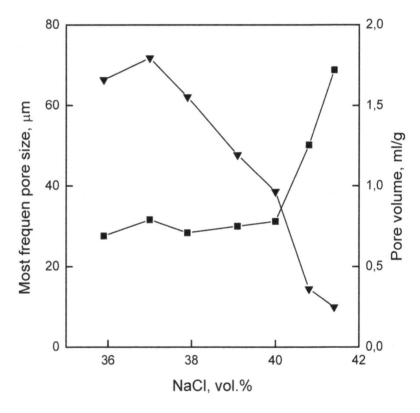

FIGURE 8 Dependence of pore volume (▼) and most frequent small superpore size (■) of PHEMA hydrogels on the content of NaCl (250–500 μm) porogen in the polymerization feed according to mercury porosimetry.

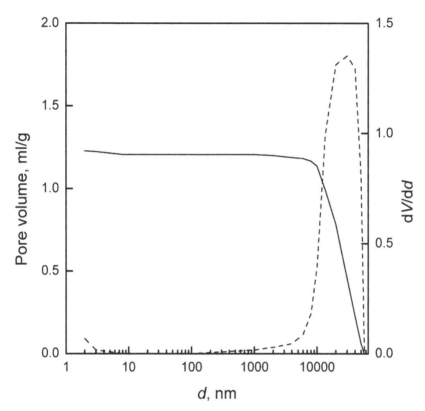

FIGURE 9 Cumulative pore volume (—) and pore size distribution (---) of the Run 4 hydrogel determined using mercury porosimetry in the range 1.88 nm–116 μm.

4.4 CONCLUSIONS

Superporous PHEMA constructs were prepared by bulk radical copolymerization of HEMA and EDMA in the presence of NaCl or/and liquid diluent (CyOH/DOH) or $(NH_4)_2SO_4$ crystals. Morphology of the prepared scaffolds was characterized by several methods including scanning electron microscopy both in swollen (LVSEM) and dry (HVSEM) state, solvent (water and cyclohexane) regains, high- and low-pressure mercury porosimetry of lyophilized samples, and dynamic desorption of nitrogen. Morphology and porous structure of the hydrogels were preferentially affected by the character and amount of the used porogen–NaCl, CyOH/DOH mixture or $(NH_4)_2SO_4$. After washing the salts and solvents from PHEMA, three types of pores were detected using microscopic and mercury porosimetry methods, including large superpores (hundreds of micrometers) as imprints of salt crystals. The hydrogels formed can be divided into two groups: disconnected and interconnected pores. The latter allowed the passage of suspension of microspheres in water, which was observed only for the samples with ammonium sulfate and the highest content of NaCl crystals used as a porogen in the feed. Interconnected pores are crucial for potential application of the scaffolds as living cell supports. LVSEM showed the undistorted (frozen) structure of the hydrogels, but only few flowed-through microspheres could be observed as they tended to escape from the pores during sample preparation. HVSEM turned out to be the best microscopic technique especially for viewing permeability of hydrogels to 8 μm microspheres. Hydrogels were initially flowed through by the particles in their natural wet state, but the specimens were then dried before SEM observation. The microparticles could be traced both on the upper/lower parts of the hydrogels and on the longitudinal sections.

Mercury porosimetry provided a detailed description of morphology of PHEMA constructs with pore sizes from units of nanometers to tens of micrometers. The drawback of the method is that the hydrogels are not measured in the swollen, but dry state, as xerogels. But comparison of the data in both wet and dry states showed that lyophilization did not change the pore structure. The mesopores and small superpores detected using mercury porosimetry cannot be formed by the imprinting mechanism. Although mesopores present only in very small amounts may be formed by phase separation, small superpores arise by polymer contraction in the walls of large superpores. Small 28–69 μm superpores were mainly present in the porous structure apart from the large superpores (200–500 μm imprints of solid porogen crystals), the volume of which was several times higher than that of other pores, as confirmed by water regain obtained by the suction method.

ACKNOWLEDGMENT

Financial support of the Grant Agency of the Czech Republic (project no. P304/11/0731) is gratefully acknowledged.

KEYWORDS

- Hydrogels
- Low-vacuum scanning electron mMicroscopy
- Poly(2-hydroxyethyl methacrylate)

REFERENCES

1. Kofron, M. D.; Cooper, J. A.; Kumbar, S. G.; and Laurencin, C. T.; Novel tubular composite matrix for bone repair. *J. Biomed Mater. Res. Part A.* **2007**, *82*, 415–425.
2. Moroni, L.; Hendriks, J. A. A.; Schotel, R.; De Wijn, J. R.; and Van Blitterswijk, C. A.; Design of biphasic polymeric 3-dimensional fiber deposited scaffolds for cartilage tissue engineering applications. *Tissue. Eng.* **2007**, *13*, 361–371.
3. Dvořánková, B.; et al. Reconstruction of epidermis by grafting of keratinocytes cultured on polymer support—clinical study. *Int. J. Dermatol.* **2003**, *42*, 219–223.
4. Bhang, S. H.; Lim, J. S.; Choi, C. Y.; Kwon, Y. K.; and Kim, B. S.; The behavior of neural stem cells on biodegradable synthetic polymers. *J. Biomater. Sci. Polym. Ed.* **2007**, *18*, 223–239.
5. Bianchi, F.; Vassalle, C.; Simonetti, M.; Vozzi, G.; Domenici, C.; and Ahluwalia, A.; Endothelial cell function on 2-D and 3-D micro-fabricated polymer scaffolds: applications in cardiovascular tissue engineering. *J. Biomater. Sci. Polym. Ed.* **2006**, *17*, 37–51.
6. Sander, E. A.; Alb, A. M.; Nauman, E. A.; Reed, W. F.; and Dee, K. C.; Solvent effects on the microstructure and properties of 75/25 poly(D,L-lactide-*co*-glycolide) tissue scaffolds. *J. Biomed. Mater. Res. Part A.* **2004**, *70*, 506–513.
7. Nam, Y. S.; Yoon, J. J.; and Park, T. G.; A novel fabrication method of macroporous biodegradable polymer scaffolds using gas foaming salt as a porogen additive. *J. Biomed. Mater. Res. Appl. Biomater.* **2000**, *53*, 1–7.
8. Kroupová, J.; Horák, D.; Pacherník, J.; Dvořák, P.; and Šlouf, M.; Functional polymer hydrogels for embryonic stem cell support. *J. Biomed. Mater. Res. Part B: Appl. Biomater.* **2006**, *76B*, 315–325.
9. Tighe, B.; and Corkhill, P.; Hydrogels in biomaterials design: is there life after poly HEMA? *Macromol. Rep.* **1994**, *A31*, 707–713.
10. Plieva, F. M.; Galaev, I. Y.; and Mattiasson, B.; Macroporous gels prepared at subzero temperatures as novel materials for chromatography of particulate-containing fluids and cell culture applications. *J. Sep. Sci.* **2007**, *30*, 1657–1671.
11. Mooney, D. J.; Baldwin, D. F.; Suh, N. P.; Vacanti, J. P.; and Langer, R.; Novel approach to fabricate porous sponges of poly(D,L-lactic-*co*-glycolic acid) without the use of organic solvents. *Biomater.* **1996**, *17*, 1417–1422.
12. Chung, H. J.; and Park, T. G.; Surface engineered and drug releasing pre-fabricated scaffolds for tissue engineering. *Adv. Drug. Deliv. Rev.* **2007**, *59*, 249–262.

13. Ferreira, L. S.; Gerecht, S.; Fuller, J.; Shieh, H. F.; Vunjak-Novakovic, G.; and Langer, R.; Bioactive hydrogel scaffolds for controllable vascular differentiation of human embryonic stem cells. *Biomater.* **2007**, *28,* 2706–2717.

14. Tangsadthakun, C.; et al. The influence of molecular weight of chitosan on the physical and biological properties of collagen/chitosan scaffolds. *J. Biomater. Sci. Polym. Ed.* **2007**, *18,* 147–163.

15. Wachiralarpphaithoon, C.; Iwasaki, Y.; and Akiyoshi, K.; Enzyme-degradable phosphorylcholine porous hydrogels cross-linked with polyphosphoesters for cell matrice. *Biomater.* **2007**, *28,* 984–993.

16. Treml, H.; Woelki, S.; and Kohler, H. H.; Theory of capillary formation in alginate gels. *Chem. Phys.* **2003**, *293,* 341–353.

17. Konno, T.; and Ishihara, K.; Temporal and spatially controllable cell encapsulation using a water-soluble phospholipid polymer with phenylboronic acid moiety. *Biomater.* **2007**, *28,* 1770–1777.

18. Heckmann, L.; et al. Human mesenchymal progenitor cell responses to a novel textured poly(L-lactide) scaffold for ligament tissue engineering. *J. Biomed. Mater. Res. Part B: Appl. Biomater.* **2007**, *81,* 82–90.

19. Darling, A. L.; and Sun, W.; 3-D microtomographic characterization of precision extruded poly-ε-caprolactone scaffolds. *J. Biomed. Mater. Res. Part B: Appl. Biomater.* **2004**, *70,* 311–317.

20. Rhee, W.; et al. In vivo stability of poly(ethylene glycol)-collagen composites. In: Poly(Ethylene Glycol) Chemistry and Biological Applications. Ed. Harris, J. M.; Zalipsky, S.; *ACS Symp. Ser.* **1997**, *680,* 420–440.

21. Savina, I. N.; Galaev, I. Y.; and Mattiasson, B.; Ion-exchange macroporous hydrophilic gel monolith with grafted polymer brushes. *J. Mol. Recognit.* **2006**, *19,* 313–321.

22. Carampin, P.; et al. Electrospun polyphosphazene nanofibers for in vitro rat endothelial cells proliferation. *J. Biomed. Mater. Res. Part A.* **2007**, *80,* 661–668.

23. Zhang, C. H.; Zhang, N.; and Wen, X. J.; Synthesis and characterization of biocompatible, degradable, light-curable, polyurethane-based elastic hydrogels. *J. Biomed. Mater. Res. Part A.* **2007**, *82,* 637–650.

24. Castner, D. G.; and Ratner, B. D.; Biomedical surface science: foundation to frontiers. *Surf. Sci.* **2002**, *500,* 28–60.

25. Lee, K. Y.; and Mooney, D. J.; Hydrogels for tissue engineering. *Chem. Rev.* **2001**, *101,* 1869–1879.

26. Refojo, M. F.; Hydrophobic interactions in poly(2-hydroxyethyl methacrylate) homogeneous hydrogel. *J. Polym. Sci. Part A1: Polym. Chem.* **1967**, *5,* 3103–8.

27. Ratner, B. D.; and Hoffman, A. S.; Hydrogels for medical and related applications. *ACS Symp. Ser.* **1976**, *31,* 1–36.

28. Horák, D.; Dvořák, P.; Hampl, A.; and Šlouf, M.; Poly(2-hydroxyethyl methacrylate-*co*-ethylene dimethacrylate) as a mouse embryonic stem cell support, *J. Appl. Polym. Sci.* **2003**, *87,* 425–432.

29. Horák, D.; Kroupová, J.; Šlouf, M.; and Dvořák, P.; Poly(2-hydroxyethyl methacrylate)-based slabs as a mouse embryonic stem cell support. *Biomater.* **2004**, *25,* 5249–5260.

30. Horák, D.; and Shapoval, P.; Reactive poly(glycidyl methacrylate) microspheres prepared by dispersion polymerization. *J. Polym. Sci. Part A: Polym. Chem. Ed.* **2000**, *38,* 3855–3863.

31. Stejskal, J.; Kratochvíl, P.; Gospodinova, N.; Terlemezyan, L.; and Mokreva, P.; Poly-aniline dispersions: preparation of spherical particles and their light-scattering characterization. *Polym.* **1992,** *33,* 4857–4858.

32. Štamberg, J.; and Ševčík, S.; Chemical transformations of polymers III. Selective hydrolysis of a copolymer of diethylene glycol methacrylate and diethylene glycol dimethacrylate. *Collect. Czech. Chem. Commun.* **1966,** *31,* 1009–2016.

33. Porosimeter Pascal 140 and Pascal 440, Instruction manual, p. 8.

34. Hradil, J.; and Horák, D.; Characterization of pore structure of PHEMA-based slabs. *React. Funct. Polym.* **2005,** *62,* 1–9.

35. Millar, J. A.; Smith, D. G.; Marr, W. E.; and Kresmann, T. R. E.; Solvent modified polymer networks. Part 1. The preparation and characterization of expanded-networks and macroporous styrene-DVB copolymers and their sulfonates. *J. Chem. Soc.* **1963,** 45, 218–225.

36. Kun, K. A.; and Kunin, R.; Macroreticular resins III. Formation of macroreticular styrene-divinylbenzene copolymers. *J. Polym. Sci. Part A1: Polym. Chem.* **1968,** *6,* 2689–2701.

37. Hradil, J.; Křiváková, M.; Starý, P.; and Čoupek, J.; Chromatographic properties of macroporous copolymers of 2-hydroxyethyl methacrylate and ethylene dimethacrylate. *J. Chromatogr.* **1973,** *79,* 99–105.

CHAPTER 5

PROGRESS ON APPLICATION OF FIRE-RETARDANT COATINGS BASED ON PERCHLOROVINYL RESIN

G. E. ZAIKOV

CONTENTS

5.1 INTRODUCTION

In many cases, polymer construction materials are a good alternative to metals and reinforced concrete. Still, it is known that the majority of such materials are combustible. That is why implementation of such materials to the building industry is associated with solving a range of engineering problems: one of them is providing the materials with the required fire safety. The fire hazard of polymers and composite materials is understood as a complex of properties that along with combustibility includes the ability for ignition, lighting, flame spreading, quantitative evaluation of smoke generation ability, and toxicity of combustion products.

Fiberglass plastics find ever-widening applications in different industrial fields. The main benefit of fiberglass plastics is higher strength and lower density compared to metals; they are not subjected to corrosion.

However, together with the valuable property complex of fiberglass plastics, they also have a significant drawback, that is, low resistance to open flame.

The sufficient increase in fire safety of fiberglass constructions may be achieved by using the passive protection measures—applying flame-retardant intumescent coatings.

Under the influence of high temperatures on the surface of an object protected from fire, an intumescent surface appears that obstructs penetration of heat and fire spread over the surface of the material.

For effective fire protection, it is necessary to apply compounds whose components inhibit combustion comprehensively: In a solid phase, it is carried out by transforming the destruction process in a material; in a gaseous phase, it is carried out by preventing the oxidation of the degradation products.[1,2]

A standard formulation of a fire-retardant coating includes an oligomer binder as also fire-retarding nitrogen, phosphorus, and/or halogen-containing inorganic and organic compounds. The fire-retarding effect is enhanced at the combination of different heteroatoms in an antipyrene.[3,4]

Previously, it was found that phosphorus–boron-containing compounds are effective antipyrenes in a fire-retardant composition.[5–7]

In the investigation, we developed a new phosphorus–boron–nitrogen-containing oligomer (PEDA). The oligomer has a good compatibility to a polymer binder, slightly migrates from a polymer material, and is an effective antipyrene when the phosphorus content is low.[6,7]

Phosphorus–boron–nitrogen-containing oligomer PEDA and polymer products comprising $—P = O$, $—P—O—B—$, $—B—O—C—$, and $—C—N—H—$ bonds are not studied enough. IR spectroscopy showed that these groups are a part of the PEDA macromolecule composition.

To improve the physical and mechanical properties of coatings and characteristics of fire protection efficiency, the fire-retardant coatings including the phosphorus–boron–nitrogen–containing oligomer PEDA as a modifier and based on perchlorovinyl resin (PVC resin) were obtained. The coatings are used for fiberglass plastics.

5.2 EXPERIMENTAL

With a purpose of defining the efficiency of the developed fire-retardant coatings for fiberglass, a set of experiments was conducted.

The experiments on fire-retardant properties were carried out on the developed technique by exposure of a coated fiberglass plastic sample to open flame. Time–temperature transformations on the nonheated surface of the fiberglass test sample was registered using a pyrometer measuring the moment of achieving the limit state—a temperature of the fiberglass destruction beginning (280–300°C).

Then, the intumescence index of the coating was calculated. The intumescence index was determined by a relative increase in height of the porous coke layer compared to the initial coating height.

The coke residue was estimated by the relative decrease in the sample weight after keeping it in the electric muffle for 10, 20, and 30 min at 600°C.

For a possibility of using the fire-retardant coatings, we need to solve a problem concerned with providing the required adhesion between the coating and protected material. The adhesive strength of the coatings to fiberglass plastic is defined as the shear strength of a joint.

During the work, the studies on the combustibility and water absorption of the coatings were conducted.

The combustibility was evaluated by exposure of a sample to the burner flame (temperature peak 840°C) and fixing the burning and smoldering time after fire source elimination.

The experiments on the water absorption were performed in distilled water at temperature $23 \pm 2°C$ for 24 hours. The water absorption was estimated by a sample weight change before and after exposure to water.

The coke microstructure formed after a test sample burning was studied as well.

5.3 RESULTS AND DISCUSSION

As part of the research, the investigation of the coatings based on perchlorovinyl resin and containing the developed intumescent additive PEDA on fire-protective properties was conducted. The results are summarized in Table 1.

TABLE 1   Influence of PEDA content on fire resistance of the coatings based on PVC resin

Parameter	Without coating	PEDA content (wt %)							
		0		2.5		5.0		7.5	
Coating thickness (mm)		0.7	1.0	0.7	1.0	0.7	1.0	0.7	1.0
Intumescence index	–	1.55	2.7	4.89	5.55	5.12	6.0	5.64	6.47
Time to the limit state (sec)	18	29	32	44	52	48	57	55	63
Temperature of the nonheated sample side in 25 seconds (°C)	–	247	223	131	115	116	108	109	102

When a coating of 1 mm in thickness containing 7.5 percent PEDA (% of the initial composition weight) is used, the peak time to the limit state is established, and the intumescence index reaches 6.47.

The temperature dependence of the nonheated sample side on flame exposure time at different PEDA content is shown in Figure 1.

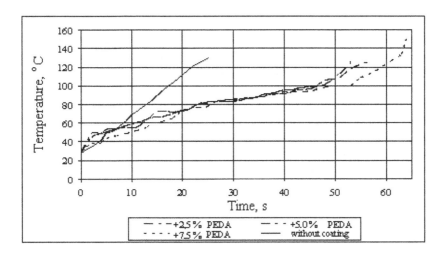

FIGURE 1   Dependence of temperature on the nonheated sample side on flame exposure time.

As shown in Figure 1, the studied coatings allow keeping temperature on the nonheated sample side within the range 80–100°C for quite a long time; time to the limit state of test samples increased by 2–2.5 times.

Coke formation is an important process in fire and heat protection of a material. Achieving a higher intumescence ratio for carbonized mass and lower heat conductivity of coke and its sufficient strength are the necessary conditions for effective fire protection.

The dependence of PEDA content influence on the ability to form coke is presented on Figure 2.

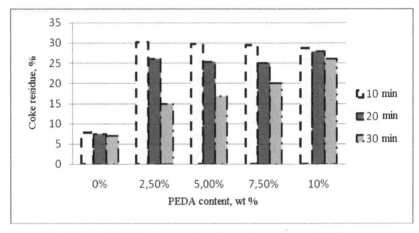

FIGURE 2   Effect of PEDA content on the coke residue values at 600°C.

As the diagram shows, with the growth of PEDA content the coke residue increases as well. This can be explained by the catalytic processes in coke formation caused by phosphorus–boron-containing substances.[8]

In the experiments on combustibility, it was found that the coatings containing PEDA are resistant to combustion and can be assigned the fire reaction class 1 as nonflammable (see Table 2).

TABLE 2   Influence of PEDA content on combustibility of the coatings based on PVC resin

PEDA content (wt %)	Combustibility of a coating
0	Burning
2.5	Self-extinguishing in 2 seconds
5.0	Self-extinguishing in a second
7.5	Not burning

The combustibility tests demonstrate that introducing PEDA into the compositions based on PVC resin promotes formation of a large coke layer: the coating film does not burn, because the presence of nitrogen in the modifying additive enables an enhancement of the fire and heat-resisting effect.

TABLE 3 Influence of PEDA content on water absorption of the coatings based on PVC resin

PEDA content (wt %)	Extent of change in sample weight	pH
0	0.02	7
2.5	−0.05	5
5.0	−0.06	5
7.5	−0.05	5
10.0	−0.07	4

The results on determining the water absorption of the modified samples revealed an insignificant washout of PEDA that takes place through the slight diffusion of the modifying additive to the film surface that is evidenced by a change in pH in 24 hours (Table 3). Nevertheless, this has no effect on fire resistance of the coatings.

As noted above, intumescent coatings should have good adhesion to the protected material; therefore, the studies on the influence of PEDA content on the adhesion strength of the coatings based on PVC resin to the fiberglass plastics were carried out while researching. The test results are illustrated in Figure 3.

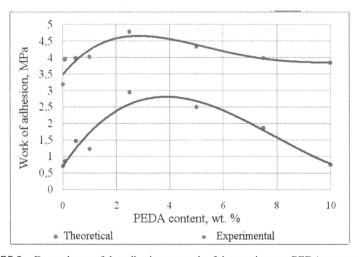

FIGURE 3 Dependence of the adhesion strength of the coatings on PEDA content.

Hence, it was established that the introduction of PEDA to the coating composition in amounts of 2.5–7.5 percent provides an increase in the adhesion strength by 1.5–4 times.

For confirmation of the experimental data, the work of adhesion was calculated according to the Young–Dupré equation. The surface tension was observed on the du Noüy tensiometer. The contact angles of wetting were defined using a goniometric method. The calculated values of the work of adhesion are well correlated with experimental data.

To improve intumescence and fire protection, the effect of introduced thermal expanded graphite (TEG), which served as a filler, on coke formation and physical and mechanical characteristics of the coatings was also studied in the work.

In a course of the study, an optimal graphite amount was chosen so that the adhesion characteristics of the coatings would not become worse and allow obtaining a sufficiently hard coke.

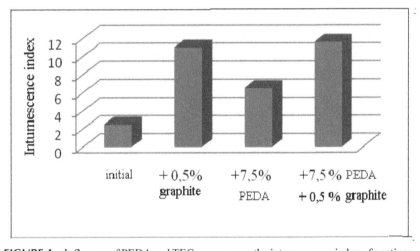

FIGURE 4   Influence of PEDA and TEG presence on the intumescence index of coatings.

The best results were achieved when applying PEDA and filler TEG. In this case, the intumescence ratio reached 11.6 (see Figure 4). The results on the influence of the coating modification and filler presence on the coke structure are presented in Figures 5–7.

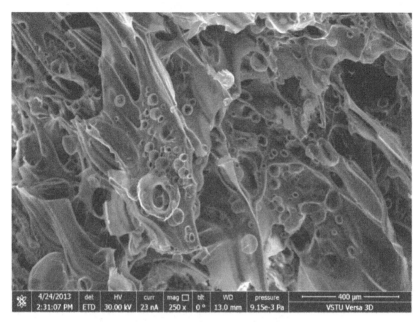

FIGURE 5   Micrograph of a coke structure on the initial coating at 250-fold magnification.

FIGURE 6   Micrograph of a coke structure on the coating containing TEG at 100-fold magnification.

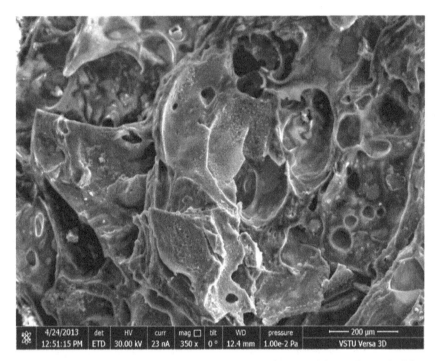

| 4/24/2013 | det | HV | curr | mag ☐ | tilt | WD | pressure | ━━━ 200 µm ━━━ |
| 12:51:15 PM | ETD | 30.00 kV | 23 nA | 350 x | 0 ° | 12.4 mm | 1.00e-2 Pa | VSTU Versa 3D |

FIGURE 7   Micrograph of a coke structure on the coating containing PEDA and TEG at 350-fold magnification.

The foamed coke formed at the testing of the composition and did not contain modifying additives and fillers had a coarse amorphous structure (Figure 5); there are foamed globular formations of 10–100 µm in the coke volume grouping to associates.

In the compositions containing TEG only (Figure 6), the coke structure is mainly determined by graphite that is present in the form of extended structures longer than 1,000 µm, 50–100 times greater than the pore size of the foamed phase. The presence of these structures leads to increased friability of the foamed mass and coke has low strength.

In the coke structure of the coating containing PEDA and TEG (see Figure 7), the extended structures, formed by graphite, disappear, and there are only short fragments of these formations. A consolidation of the carbon layers is observed, which, probably, takes place due to formation of the high temperatures of polyphosphoric acids on the surface and between the layers of expanded graphite sites that solder layers and, thereby, impede TEG intumescence; there is a slight shrinkage of the intumescent layer. As a result, the intumescence index of this composition does not substantially exceed the intumescence index of the composition containing only filler, but with a more ordered structure of coke and a sufficiently high strength and hardness of the composition the fire resistance increases. Such a coke can withstand more intense combustion gas streams.

5.4 CONCLUSION

Thus, the fire-retardant coatings based on the developed phosphorus–boron–nitrogen-containing oligomer have high fire and heat protective and adhesive properties. The structure and presence of phosphorus, boron, and nitrogen heteroatoms promotes an enhancement of the film-forming polymer carbonization and increase in the intumescence ration of the coatings. In addition, the definite advantage of PEDA application is that it is slightly washed out of a coating when exposed to water.

Introduction of the modifying additive PEDA in combination with a filler—TEG—permits to increase the coating intumescence by 11 times, resulting in improved fire and heat protective properties of the coatings and reduced destruction of fiberglass plastic.

The research was conducted with financial support from the Ministry of Education and Science of the Russian Federation under realization of the federal special-purpose program scientific, academic, and teaching staff of innovative Russia for 2009–2013 years: the Grant Agreement No. 14.B37.21.0837 "Development of active adhesive compositions based on element organic polymers and vinyl monomers."

KEYWORDS

- **Fire-retardant coatings**
- **Perchlorovinyl resin**
- **Phosphorus–boron–nitrogen-containing oligomer (PEDA)**

REFERENCES

1. Berlin, Al. Al.; Combustion of polymers and polymer materials of reduced combustibility. *Soros Educ. J.* **1996,** *9,* 57–63.
2. *Shuklin, S. G.; Kodolov, V. I.; and Klimenko, E. N.; Intumescent coatings and the processes that take place in them. Fiber Chem.* **2004,** *36(3),* 200–205.
3. Nenakhov, S. A.; and Pimenova, V. P.; Physical transformations in fire retardant intumescent coatings based on organic and inorganic compounds. *Pozharovzryvobezopasnost – Fire Explos. Saf.* **2011,** *20(8),* **17–24.**
4. Balakin, V. M.; and Polishchuk, E. Yu.; Nitrogen and phosphorus containing antipyrenes for wood and wood composite materials. *Pozharovzryvobezopasnost – Fire Explos. Saf.* **2008,** *17(2),* 43–51.
5. Shipovskiy, I. Ya.; Bondarenko, S. N.; and Goryainov, I. Yu.; Fire protective modification of wood. Proceedings of the International Scientific and Practical Conference. *Dnepropetrovsk,* **2005,** *47,* 20.
6. Gonoshilov, D. G.; Keibal, N. A.; Bondarenko, S. N.; and Kablov, V. F.; Phosphorus boron containing fire retardant compounds for polyamide fibers. Proceedings of the 16th International Scientific and Practical Conference. Rubber Industry: Raw Materials. *Manuf. Mater. Technol.* Moscow, **2010,** 160–162.
7. Lobanova, M. S.; Kablov, V. F.; Keibal, N. A.; and Bondarenko, S. N.; Development of active adhesive fire and heat retardant coatings for fiber-glass plastics. All the Materials. Encyclopaedic Reference Book. **2013,** *04,* 55–58.
8. Korobeynichev, O. P.; Shmakov, A. G.; and Shvartsberg, V. M.; The combustion chemistry of organophosphorus compounds. *Uspekhi. Khimii.* **2007,** *76(11),* 1094–1121.

CHAPTER 6

THE DEPENDENCIES OF DISSOCIATION ENERGY OF BINARY MOLECULES AND FORMATION ENTHALPIES OF SINGLE-ATOM GASES ON INITIAL SPATIAL-ENERGY CHARACTERISTICS OF FREE ATOMS

G. A. KORABLEV, N. G. KORABLEVA, and G. E. ZAIKOV

CONTENTS

6.1   INTRODUCTION

Thermodynamic parameters (enthalpy, entropy, thermodynamic potential) allow the description of various physical, chemical, and other processes and explicitly assess the possibility of their flow without vividly using physical models. The application of reliable values of formation enthalpies is required for searching new perspective materials and compounds, assessment of molecule kinetic properties, analysis mechanisms of rocket propellant combustion, etc. [1–3].

The establishment of connection between the structure and thermochemical parameters, as well as kinetic and thermochemical characteristics of interacting systems is of great importance.[4,5] But the analysis of dependencies between the main parameters of chemical thermodynamics and spatial-energy characteristics of free atoms is still topical.

For this purpose, the methodology of spatial-energy parameter (P-parameter) has been used in this research.[6]

6.2   RESEARCH TECHNIQUE

The comparison of multiple regularities of physical and chemical processes allows the assumption that in many cases the principle of adding reciprocals of volume energies or kinetic parameters of interacting structures is fulfilled.

Some examples include ambipolar diffusion, cumulative rate of topochemical reaction, and change in the light velocity when moving from vacuum into the given medium and resultant constant of chemical reaction rate (initial product–intermediary activated complex–final product).

Lagrangian equation for the relative motion of isolated system of two interacting material points with masses m_1 and m_2 in the coordinate x with acceleration α can be as follows:

$$\frac{1}{1/(m_1 a\Delta x)+1/(m_2 a\Delta x)} \approx -\Delta U \text{ or } \frac{1}{\Delta U} \approx \frac{1}{\Delta U_1}+\frac{1}{\Delta U_2} \tag{1}$$

where ΔU_1 and ΔU_2—potential energies of material points on elemental interaction section and ΔU—resultant (mutual) potential energy of these interactions.

The atomic system is formed by oppositely charged masses of nucleus and electrons. In this system, the energy characteristics of the subsystems include orbital energy of electrons and effective energy of nucleus considering screening effects (by Clementi). Either the bond energy of electrons (W) or the ionization energy of an atom (E_i) can be used as the orbital energy.

Therefore, assuming that the resultant interaction energy in the system's orbital nucleus (responsible for interatomic interactions) can be calculated following the principle of adding reciprocals of some initial energy components, the introduction of P-parameter [6,7] as averaged energy characteristic of valence orbitals based on the following equations has been substantiated:

$$\frac{1}{q^2/r_i} + \frac{1}{W_i n_i} = \frac{1}{P_э} \tag{2}$$

$$P_E = \frac{P_0}{r_i} \tag{3}$$

$$\frac{1}{P_0} = \frac{1}{q^2} + \frac{1}{(wrn)_i} \tag{4}$$

$$q = \frac{Z^*}{n^*} \tag{5}$$

where W_i—bond energy of an electron [8]; n_i—number of elements of the given orbital; r_i—orbital radius of i—orbital [9]; and Z^* and n^*—effective charge of a nucleus and effective main quantum number.[10,11]

The value of P_0 is called spatial-energy parameter and the value of P_E—effective P-parameter. The effective P_E-parameter has a physical sense of some averaged energy of valent electrons inside the atom and is measured in energy units, for example, in electron volts (eV).

The calculations demonstrated that the values of P_E-parameters numerically equal (in the limits of 2%) the total energy of valent electrons (U) by statistic model of an atom. Using the known relationship between the electron density (β) and interatomic potential by statistic model of an atom, the direct dependence of P_E-parameter on the electron density on the distance r_i from the nucleus can be obtained:

$$\beta_i^{2/3} = \frac{AP_0}{r_i} = AP_E$$

where A—constant.

Based on Eqs (2–5), P_E and P_0-parameters of free atoms for the majority of elements of the periodic system have been calculated.[6,7] Some of these calculations are summarized in Table 1.

Modifying the rules of addition for reciprocals of energy values of sub-systems with reference to complex structures, the formula for calculating P_C-parameter of a complex structure is obtained as follows:

$$\frac{1}{P_c} = \left(\frac{1}{NP_E} \right)_1 + \left(\frac{1}{NP_E} \right)_2 + L \tag{6}$$

where N_1 and N_2—number of homogeneous atoms in subsystems.

6.3. CALCULATION OF DISSOCIATION ENERGY OF BINARY MOLECULES VIA AVERAGE VALUES OF P_0-PARAMETERS

The application of methods of valent bond and molecular orbitals to complex structures faces significant obstacles for prognosticating energy characteristics of the bonds formed. Based on Eq. (6), the application of the formula for calculating the dissociation energy has turned out to be practically possible [12]:

$$\frac{1}{D_0} = \frac{1}{P_C} = \frac{1}{\left(P_E \frac{N}{K} \right)_1} + \frac{1}{\left(P_E \frac{N}{K} \right)_2} \tag{7}$$

where N—bond order, K—maxing or hybridization coefficient that usually equals the number of valent electrons considered, and the value of $P_E(N/K)$ has a physical sense of the averaged energy of spatial-energy parameter accrued to one valent electron of the orbitals registered. For complex structures, P_E-parameter was averaged by all main valent orbitals.

For binary molecules, the dissociation energy (D_0) corresponds to the value of chemical bond energy: $D_0 = E$.

The calculation results of dissociation energy in Eq. (7), summarized in Table 2, demonstrated that $P_C = Д_0$. For some molecules containing such elements as F, N, and O, the values of ion radius have been applied to register the bond ionic character for calculating P_E-parameter (in Table 2 marked with *). For such molecules as C_2, N_2, and O_2, the calculations have been made by multiple bonds. In other cases, the average values of bond energy have been applied. The calculated data do not contradict the experimental data.[2,3]

6.4 EQUATION OF DEPENDENCE OF THERMODYNAMIC CHARACTERISTICS UPON SPATIAL-ENERGY PARAMETERS OF FREE ATOMS

The dissociation energy (E) of the molecular breakage into two parts numerically equals the difference of the heat produced during the formation of dissociation products and during the initial molecule formation:

$$E = D_0 = [\Delta H_0(R_1) + \Delta H_0(R2)] - \Delta H_0(R_1R_2) \tag{8}$$

where D_0—dissociation energy of a molecule (rupture energy of its bond), $\Delta H_0(R_1)$ and $\Delta H_0(R_1)$—formation enthalpy at $0°K$ of dissociation products R_1 and R_2, respectively, $\Delta H_0(R_1\,R_1)$—initial energy formation enthalpy.

Because for binary molecules $E = D_0$, Eq. (7) gives

$$E = \frac{\left(P_E\dfrac{N}{K}\right)_1\left(P_E\dfrac{N}{K}\right)_2}{\left(P_E\dfrac{N}{K}\right)_1 + \left(P_E\dfrac{N}{K}\right)_2} \tag{8a}$$

From Eqs. (8) and (7a)

$$\frac{\left(P_E\dfrac{N}{K}\right)_1\left(P_E\dfrac{N}{K}\right)_2}{\left(P_E\dfrac{N}{K}\right)_1 + \left(P_E\dfrac{N}{K}\right)_2} = \Delta H_0(R_1) + \Delta H_0(R_2) - \Delta H_0(R_1R_2) \tag{9}$$

This is the equation of direct dependence of thermodynamic values and initial spatial-energy characteristics of a free atom.

The relationship obtained (9) allows defining the corresponding dependencies with the experimental thermodynamic characteristics of chemical reactions.

Thus, in thermodynamic experimental methods for determining the formation enthalpy, the following equation is used:

$$\Delta H_0 = T\left(\Delta \Phi_T^* - R \ln K_E\right)$$

where K_E—equilibrium constants of chemical reaction; T—thermodynamic temperature of the process; $\Delta \Phi_T^*$—change in the reaction of considered reduced thermodynamic potential; R—universal gas constant.

6.5   EVALUATION OF SINGLE-ATOM GAS FORMATION ENTHALPY (ΔH_G^0)

In the experimental evaluation method of single-atom gas formation enthalpy from solids, a known thermodynamic identity can be used:

$$\Delta H_G^{\,0} \equiv \Delta H^0_{SD} + \Delta H_{S,298} \tag{10}$$

where $\Delta H_G^{\,0}$—formation enthalpy of gaseous substance; ΔH^0_{SD}—formation enthalpy of solid in nonstandard state; and $\Delta H_{S,298}$—sublimation enthalpy (substance transition from solid into gaseous state).

The heats of formation of chemical elements in standard state are considered to be zero. The list of such elements is known, for example, [13] and mainly incorporate elements in solid state. Thus, the single-atom gas formation enthalpy in these cases equals the sublimation enthalpy and is determined by physical and chemical criteria of the process. The sublimation comes to diffusion movements of heated particles into the surface layer with further effusion (outflow). The diffusion activation energies in external and internal regions are quite different.[14]

It is usually considered that sublimation flows are due to the particle migration from more strongly bonded state with the highest number of neighbors to less strongly bonded state and further to the adsorbed surface layer.[15] By analogy, the number of interacting molecules in liquid decrease, when molecules move from the lower part of the surface layer to its upper part.

The diffusion activation energy is defined by the values of electron densities of migrating particle and particles surrounding it at the distance of atom–molecule interaction radius (R), that is, such interaction is carried out through the power field of particles evaluated by the values of their P-parameters.[16] In such a model, the diffusion process has a similar nature in any aggregate state and its energy and directedness are defined by the following three main factors:

(1) value of P-parameters of structures

(2) number of particles

(3) radius of atom–molecule interaction

In advance researches carried out for solid solutions in the frames of generalized grid model, it has been found out that "effective diffusion coefficient depends on local composition, own volumes of component atoms, and potentials of paired interactions." [17] At the same time two types of diffusion are also distinguished: "normal" and "bottom-up" [18]: this apparently agrees with the notion of volume (internal) and surface diffusion.

TABLE 1 P-parameters of atoms calculated through bond energy of electrons

Atom	Valent elec-trons	W (eV)	r_i (Å)	q^2_0 (eVÅ)	P_0 (eVÅ)	R (Å)	P_0/R (eV)	$P_0/$ $R(n^* + 1)$	r_1 (Å)	P_0/r_1 (eV)	$P_0/$ $r_1(n^* + 1)$
1	2	3	4	5	6	7	8	9	10	11	12
Li	$2s^1$	5.3416	1.506	5.86902	3.475	1.55	2.2419	0.7473	0.68	5.1103	1.7034
Na	$3s^1$	4.9552	1.713	10.058	4.6034	1.69	2.4357	0.60892	0.96	4.6973	1.1743
K	$4s^1$	4.0130	2.612	10.993	4.8490	2.36	2.0547	0.4372	1.33	3.6459	0.7757
Rb	$5s^1$	3.7511	2.287	4.309	5.3630	2.48	2.1625	0.43250	1.49	3.5993	0.71986
Cs	$6s^1$	3.3647	2.516	16.193	5.5628	2.68	2.0757	0.3992	1.65	3.3714	0.6483
Mg	$3s^1$	6.8859	1.279	17.501	5.8568	1.60	3.6616	0.91544	0.74	7.9173	1.9793
	$3s^2$	6.8859	1.279	17.501	8.7787	1.60	514667	1.3717	0.74	11.863	2.9658

TABLE 1   *(Continued)*

Atom	Valent electrons	W (eV)	r_1 (Å)	q_0^2 (eVÅ)	P_0 (eVÅ)	R (Å)	P_0/R (eV)	$P_0/R(n^*+1)$	r_1 (Å)	P_0/r_1 (eV)	$P_0/r_1(n^*+1)$
Ca	$4s^1$	5.3212	1.690	17.406	5.929	1.97	3.0096	0.64035	1.33		
	$4s^2$	5.3212	1.690	17.406	6.6456	1.97	4.902	0.95535	1.04	8.5054	1.8097
Sr	$5s^1$	4.8559	1.836	21.224	6.790	2.15	2.9205	0.56409			
	$5s^2$				9.6901	2.15	4.5070	0.90140	1.20	8.0751	1.6150
Ba	$6s^1$	4.2872	2.060	22.950	6.3768	2.21	2.6854	0.5549	1.38	7.2328	1.3909
	$6s^2$				9.9812	2.21	4.5164	0.8685			
1	2	3	4	5	6	7	8	9	10	11	12
	$4s^1$	5.7174	1.570	19.311	6.1279	1.64	3.7365	0.7950			
Sc	$4s^2$	5.7174	1.570	19.311	9.3035						
	$3d^1$	9.3532	0.539	81.099	4.7463						
	$3d^1$	9.3532	0.539	81.099	14.050	1.64	2.8941	0.6158			
	$4s^2+3d^1$					1.64	8.5671	1.8228	0.83	16.928	3.6016
	$5s^1$	6.3376	1.693	22.540	6.4505	1.81	3.5638	0.71276			
Y	$5s^2$				10.030		5.5417	1.1083			
					5.6756		3.1357	0.62714			
	$4d^1$	6.7965	0.856	229.18	15.706		8.6771	1.7354			
	$4d^1+5s^2$								0.97	16.192	3.2384
	$6s^1$	4.3528	1.915	34.681	6.7203	1.87	3.5937	0.6911			
	$6s^2$				11.259		6.0209	1.1579			
								0.4378			
	$4f^1$	10.302	0.4234	17870	4.2576		2.2768	1.8922			
La	$5p^1$	25.470	0.827	145.53	18.400		9.8396	3.0501			
	$6s^2+5p^1$				29.659		15.860	1.5957	1.04	28.518	5.4843
	$6s^2+4f^1$				15.517		8.2976		1.04	14.920	2.8693
	$4s^1$	6.0082	1.477	20.879	6.2273	1.46	4.2653	0.9075			
	$4s^2$	6.0082	1.477	20.879	9.5934	1.46	6.5708				
		11.990	0.496	106.04	5.556	1.46	3.8053	1.3960	0.78	12.299	2.6169
Ti	$3d^1$	11.990	0.469	106.04	10.558			0.80965			
	$3d^2$				20.151						
	$4s^2+3d^2$					1.46	13.802	2.9366	0.64	31.486	6.6981
	$4s^1+3d^1$							1.7172			

TABLE 1 *(Continued)*

Atom	Valent elec-trons	W (eV)	r_i (Å)	q^2_0 (eVÅ)	P_0 (eVÅ)	R (Å)	P_0/R (eV)	$P_0/$ R(n*+1)	r_i (Å)	P_0/r_i (eV)	$P_0/$ r_i(n* +1)
	5s¹	5.5414	1.593	23.926	6.5330	1.60	4.0831	0.8166			
	5s²				10.263		6.4146	1.2829			
					6.9121		4.3201	0.8640			
Zr	4d¹	9.1611	0.790	153.76	13.229		8.2681	1.6536			
	4d²5s² + 4d²				23.492		14.683	2.9365			
						1.6			0.82	28.640	5.7298

TABLE 2 Dissociation energies of two-atom molecules—$D_0 \left(\frac{KJ}{mol}\right)$

Struc-ture	First atom				Second atom				P_C (eV)	D_0 calcul.	D_0 experim.
	Orbitals	N/k	P_E (eV)	$P_E \dfrac{N}{k}$	Orbitals	N/k	P_E (eV)	$P_E \dfrac{N}{k}$			
1	2	3	4	5	6	7	8	9	10	11	12
CCl	2p¹	1/1	7.6208	7.6208	3P¹	1/1	8.5461	8.5461	4.0209	388.9	393.3
CBr	2p¹	1/1	7.6208	7.6208	4P¹	1/1	8.0430	8.0430	3.9130	377.7	364
CJ	2p¹	1/1	7.6208	7.6208	5P¹	1/1	7.2545	7.2545	2.2523	217.4	209.2
CN	2p²	2/2	13.066	13.066	2S²2P³	2/5	47.413	18.965	7.7358	746.7	755.6
CN	2p²	2/2	14.581	14.581	2P³	2/3	25.127	16.751	7.796	752.5	755.6
C-O	2p²	1/2	13.066	6.533	2P²	1/2	17.967	8.984	3.782	365	356
NO	2p¹	1/1	9.2839	9.2839	2P²	2/2	20.048	20.048	6.346	612.5	626.8
CH	2p²	1/2	13.066	6.533	1S¹	1/1	9.066	9.066	3.7969	366.5	333±1
OH	2p²	1/2	17.967	8.9835	1S¹	1/1	9.066	9.066	4.5118	435.5	423.7
ClF	3s²3p⁵	1/7	29.391*	4.1987	2S²2P⁵	1/7	38.202*	5.4574	2.5579	246.9	229.1
ClO	3s²3p⁵	1/7	29.391*	4.1987	2P²	2/2	8.7191*	8.7191	2.8337	273.5	264
ClO	3p¹	1/1	4.7216*	4.7216	2S²2P⁴	1/6	30.738*	5.123	2.450	237.2	264
FO	2p¹	1/1	4.9887*	4.9887	2P²	1/2	8.7191*	4.3596	2.327	224.6	219.2
NF	2p³	1/3	10.696*	3.5653	2P¹	1/7	38.202*	5.4574	2.486	239.5	298.9
NCl	2p³	1/3	22.296	7.432	3P¹	1/1	8.5461	8.5461	3.9751	383.7	384.9
H₂	1s¹	1/1	9.0624	9.0624	1S¹	1/1	9.066	9.066	4.533	437.5	432.2
Li₂	2s¹	1/1	2.2419	2.2419	2S¹	1/1	2.2419	2.2419	1.121	108.2	98.99

TABLE 2   *(Continued)*

Structure	First atom				Second atom				P_C (eV)	D_0 calcul.	D_0 experim.
	Orbitals	N/k	P_E (eV)	$P_E \dfrac{N}{k}$	Orbitals	N/k	P_E (eV)	$P_E \dfrac{N}{k}$			
B_2	$2p^1$	1/1	5.4885	5.4885	$2P^1$	1/1	5.4885	5.4885	2.744	264.9	276±21
C-C	$2p^1$	1/1	7.6208	7.6208	$2P^1$	1/1	7.6208	7.6208	3.810	367.8	376.7
C=C	$2p^2$	2/2	13.066	13.066	$2P^2$	2/2	13.066	13.066	6.533	630.6	611
N-N	$2p^3$	1/3	10.696*	3.5653	$2P^3$	1/3	10.696*	3.5653	1.783	172.1	161
N=N	$2s^2 2p^3$	2/5	22.745*	9.098	$2S^2 2P^3$	2/5	22.745*	9.098	4.549	439	418
O-O	$2p^2$	1/2	8.7191	4.3596	$2P^2$	1/2	8.7191	4.3596	2.1798	210.4	213.4
O=O	$2s^2 2p^4$	2/6	30.738*	10.246	$2S^2 2P^4$	2/6	30.738*	10.246	5.123	494.5	498.3

In a liquid, the radius of the sphere of molecular interaction $R \approx 3r$, where r—radius of a molecule. Liquids are mainly formed by the elements of first and second periods of the system. For the second period, we can write: thus $R \approx 3r = (n + 1)r$, where n—main quantum number. For both periods (first and second), $R = (<n> + 1)r \approx 2.5r$.

Let us assume that this principle with definite approximation can be extended to various aggregate states of elements of all other periods but taking into consideration screening effects introducing the value of effective main quantum number (n^*) instead of n. These values of n^* and $n^* + 1$ used by Slater [19] are summarized in Table 3.

TABLE 3   Effective main quantum number

n	1	2	3	4	5	6
n^*	1	2	3	3.7	4	4.2
$n^* + 1$	2	3	4	4.7	5	5.2

Thus, we assume that the radius of sphere of atom–molecule interaction during the particle diffusion in the sublimation process is defined as follows:

$$R = (n^* + 1)r$$

where r—dimensional characteristic of atom structure. Total change of R is from $3r$ to $5.2r$.

When forming single-atom gases, the sublimation process is accompanied by the rupture of paired bond of atoms of nearby surroundings. The averaged value of structural P_S-parameter of interacting atoms can be the assessment of formation enthalpy and numerically equals the value of P_0-parameter falling on the radius unit of atom–molecule interaction but taking into consideration the relative number of interacting particles by the equation:

$$P_S = \frac{P_0}{R}\gamma = \frac{P_0\gamma}{r(n^* + 1)} \approx \Delta H_\Gamma^\circ$$

where γ—coefficient taking into consideration the relative number of interacting particles and equaled to (as the calculations demonstrated):

$$\gamma = \frac{N_0}{N}$$

where N_0—number of particles in the sphere volume of the radius R, N—number of particles of realized interactions depending on the process type (internal or surface diffusion).

Inside the liquid below the top layer $2R$ thick, the resultant force of molecular interaction equals zero.

Applying the initial analogy to internal diffusion and sublimation, we can consider that such equilibrium state corresponds to the equality $N_0 = N$ and then $\gamma = 1$.

On the top part of liquid surface layer, the volume of the sphere of atom–molecule interaction and the number of particles in it is practically two times lower in comparison with internal layers under $2R$, that is, $\frac{N_0}{N} \approx \frac{1}{2}$ and $\gamma = \frac{1}{2}$—for surface diffusion and sublimation.

In case of volume diffusion, even a more extreme option is possible, where the number of particles of realized interactions exceeds N_0 by two times.

Thus, two-valent elements magnesium and calcium form either two or even four covalent bonds (in chelate compounds): two covalent bonds and two by donor–acceptor mechanism. For such and analogous cases with volume diffusion $\gamma = 2$.

6.6 CALCULATIONS AND COMPARISONS

Based on such initial statements and assumptions, the value of P_S-parameter was calculated (and ΔH^0_r) using Eq. (12) for the majority of elements in periodic system (Table 4). The values of covalent, atom, and ion radii were calculated by Belov–Bokiy and partially by Batsanov [20].

The structural coefficient γ was 2 for covalent spatial-energy bonds at the distances of only atom or covalent radii (surface diffusion). Coefficient γ was 1 for ion spatial-energy bonds of elements from subgroups a and group 8 in periodic system (internal diffusion). In all other cases, this coefficient equaled 1 or 1/2 (internal or volume diffusion).

In some cases, the values of ΔH^0_r of the given element was found to be average because its two possible valent states (marked with <...>). The calculations are carried out practically for all elements of six periods irrespective of their aggregate state that definitely could give higher accuracy than reference data.

Experimental and reference data [2,3] have a relative error within 0.5–1.5 (%), but in the given calculations such average error was about 5 percent. Probably the search for a more rational technique of registering screening effects for clarifying the effective main quantum number can eventually bring more reliable results for evaluating ΔH^0_r.

6.7 CONCLUSIONS

The dependencies of dissociation energy of binary molecules and formation enthalpies of single-atom gases upon initial spatial-energy characteristics of free atoms have been determined.

The corresponding equations, the calculations that agree with the experimental and reference data, have been obtained.

TABLE 4 Calculation of formation enthalpy of single-atom gases — ΔH_{298} 1 eV = 96.525 kJ/mol

Group	Atom	Orbitals	P_0 (eVÅ)	γ	r (Å)	$\Delta H = \dfrac{P_0\gamma}{r(n^*+1)}$ (eV)	r_H (Å)	$\Delta H = \dfrac{P_0\gamma}{r_H(n^*+1)}$ (eV)	ΔH calculation $\left(kJ/mol\right)$	ΔH^0_{298} ref. data $\left(kJ/mol\right)$
1	2	3	4	5	6	7	8	9	10	11
	Li	$2s^1$	3.4750	1			0.68	1.703	164.4	159.3
	Na	$3s^1$	4.6034	1			0.98	1.1743	113.3	107.5
1a	K	$4s^1$	4.8490	2	2.36	0.874			84.36	88.9
	Rb	$5s^1$	5.3630	2	2.48	0.865			83.49	80.9
	Cs	$6s^1$	5.5628	2	2.68	0.7984			77.07	77
	Be	$2s^2$	7.512	½			0.34	3.682	355.4	326.4
	Mg	$3s^2$	8.7787	½			0.74	1.483	143.1	147.1
2a	Ca	$4s^2$	8.8456	1			1.04	1.8097	174.7	177.8
	Sr	$5s^2$	9.6901	1			1.20	1.6150	155.9	160.7
	Ba	$6s^2$	9.9812	2	2.21	1.737			167.7	179.1
	Sc	$4s^2 3d^1$	14.050	1			0.74	4.040	389.96	379.1
3a	Y	$5s^1 4d^2$	17.527	2	1.62	4.342			419.2	423
	La	$6s^2 5p^1$	29.659	2	1.87	6.1002			<448.4>	429.7
		$6s^2 4f^1$	15.517	2	1.87	3.1915				

TABLE 4   *(Continued)*

Group	Atom	Orbitals	P_0 (eVÅ)	γ	r (Å)	Δ $H = \dfrac{P_0\gamma}{r(n^*+1)}$ (eV)	r_H (Å)	$\Delta H = \dfrac{P_0\gamma}{r_H(n^*+1)}$ (eV)	ΔH calculation $\left(\text{kJ}/\text{mol}\right)$	ΔH^0 298 ref. data $\left(\text{kJ}/\text{mol}\right)$
1	2	3	4	5	6	7	8	9	10	11
	Ti	$4s^1\,3d^1$	11.7853	2	1.46	3.4344			<449.2>	468.6
		$4s^2\,3d^2$	20.151	2	1.46	5.8732				
4a	Ti	$4s^1\,3d^{1^1}$	11.7853	1			0.78	3.214	<478.4>	468.6
		$4s^2\,3d^2$	20.151	1			0.64	6.6981		
	Zr	$5s^2\,4d^2$	23.492	2	1.60	5.873			566.9	600
	Hf	$6s^2\,5d^2$	24.498	2	$r_H =$ 1.44	6.543			631.6	620.1
	V	$4s^2\,3d^1$	15.776	1			0.67	5.6010	<512.5>	514.6
		$4s^1\,3d^2$	17.665	1			0.67	5.6090		
	Nb $(5s^2\,4d^3)$	$5s^2\,4d^2$	25.577	1			0.67	7.6349	736.9	722.6
5a	Nb $(5s^1\,4d^4)$	$5s^1\,4d^2$ $5s^1\,4d^4$	20.805 30.607	1 1			0.767 0.65	5.425 9.2748	<709.4>	722.6
	Nb (3, 4, 5)			1					<723.2>	722.6
	Ta	$6s^2\,5d^2$ $6s^2\,5d^3$	26.314 32.722	1 1			0.737 (0.66)	6.8662 9.5344	<791.5>	786.1

TABLE 4   *(Continued)*

Group	Atom	Orbitals	P_0 (eVÅ)	γ	r (Å)	$\Delta H = \dfrac{P_0\gamma}{r(n^*+1)}$ (eV)	r_{H} (Å)	$\Delta H = \dfrac{P_0\gamma}{r_{\mathrm{H}}(n^*+1)}$ (eV)	ΔH calculation $\left(\mathrm{kJ/mol}\right)$	ΔH^0 298 ref. data $\left(\mathrm{kJ/mol}\right)$
	Cr	$4s^2$	10.535	1			0.83	2.7006	<406.1>	397.5
		$4s^2\,3d^1$	17.168	1			0.64	5.7074		
6a	Mo (2)	$5s^1\,4d^1$ $(5s^2\,4d^4)$	19.574	1			0.737	5.312	<658>	656.5
	Mo (4)	$(5s^1\,4d^4)\;5s^1\,4d^3$	28.293	1			0.68	8.3215		
1	2	3	4	5	6	7	8	9	10	11
	W (4)	$6s^2\,6s^1$	27.879	2	1.40	3.8295			<831.5>	856.9
	W (5)	$6s^2\,5d^3$	34.828		1.40	4.7841				
	Mn (2)	$4s^1\,3d^1$	12.924	1			0.91	3.0217	291.7	284.5
7a	Tc (3)	$5s^1\,4d^2$	23.866	2	1.36	7.0194			674.1	657
	Re (4)	$6s^2\,5d^2$	29.806	1			0.72	7.961	768.4	775.7
	Re (6)	$6s^2\,5d^4$	44.519	$\frac{1}{2}$			0.52	8.232	794.6	
	Re (4, 6)								<781>	

TABLE 4   *(Continued)*

Group	Atom	Orbitals	P_0 (eVÅ)	γ	r (Å)	$\Delta H = \dfrac{P_0\gamma}{r(n^*+1)}$ (eV)	r_H (Å)	$\Delta H = \dfrac{P_0\gamma}{r_H(n^*+1)}$ (eV)	ΔH calculation $\left(\text{kJ}/\text{mol}\right)$	ΔH^0 298 ref. data $\left(\text{kJ}/\text{mol}\right)$
	Fe (2)	$4s^1 3d^1$	12.717	1			0.80	3.3822	<418.6>	417.1
	Fe (3)	$4s^2 3d^1$	16.664	1			0.67	5.2918		
	Co (2)	$4s^1 3d^1$	12.707	1			0.78	3.4662	<436.5>	428.4
	Co (3)	$4s^2 3d^1$	16.680	1			0.64	5.5785		
8	Ni (2)	$4s^1 3d^1$	12.705	1			0.74	3.6530	<465.5>	429.3
	Ni (3)	$4s^1 3d^2$	16.897	1			0.60	5.992		
	Ni (2)	$4s^1 3d^1$	12.705	2	1.24	4.360			420.8	429.3
	Ru (3)	$5s^1 4d^2$	23.636	1			0.68	6.982	671	656.8
	Rh (3)	$5s^2 4d^1$	21.114	1			0.75	5.6304	543.5	556.5
	Pd (2)	$5s^2$	12.057	1			0.64	3.7681	363.7	372.3
	Os (3)	$5d^3$	25.986	2	$r^k = 1.26$	7.922			766	790.2
1	2	3	4	5	6	7	8	9	10	11
	Ir (2)	$6s^1 5d^1$	17.631	2	1.35	5.023			<665.8>	669.5
8	Ir (4)	$6s^2 5d^2$	30.790	2	1.35	8.772				
	Pt (2)	$6s^1 5d^1$	17.381	1			0.6	5.971	576.4	565.7

TABLE 4 *(Continued)*

Group	Atom Orbitals	P_0 (eVÅ)	γ	r (Å)	$\Delta H = \dfrac{P_0\gamma}{r(n^*+1)}$ (eV)	r_{H} (Å)	$\Delta H = \dfrac{P_0\gamma}{r_{\text{H}}(n^*+1)}$ (eV)	ΔH calcula-tion $\left(\text{kJ}/\text{mol}\right)$	ΔH^0 298 ref. data $\left(\text{kJ}/\text{mol}\right)$
	Cu (2) $4s^1 3d^1$	13.242	1			0.80	3.5218	339.9	337.6
1_6	Ag $4d^1$	9.7843	2		3.131			<282.3>	284.9
	Ag (1) $4d^1$	9.7843	2	$r^k =$ 1.25 1.44	2.718				
	Au (3) $6s^1 5d^2$	27.536	1	1.44	3.6774			355.0	368.8
	Zn (1) $3d^1$	6.1153	1	1.39	0.9361			<127.1>	130.5
	Zn (2) $4s^2$	11.085	1	1.39	1.6968				
2_6	Cd (2) $5s^2$	11.839	$\frac{1}{2}$			0.99	1.1959	115.4	111.8
	Cd (2) $5s^1 4d^1$	17.145	$\frac{1}{2}$	1.56	1.0991			106.1	111.8
	Cd (1, 2)							<110.8>	111.8
	Hg (1) $5d^1$	11.266	$\frac{1}{2}$	1.60	0.6771			65.35	61.40
	B $2s^2 2p^1$	16.086	1	0.91	5.892			568.7	561
3_6	Al $3s^2 3p^1$	18.093	1	1.43 $r^k =$ 1.25	3.163 3.619			<327.3>	329.3

TABLE 4   *(Continued)*

Group	Atom	Orbitals	P_0 (eVÅ)	γ	r (Å)	$\Delta H = \dfrac{P_0\gamma}{r(n^*+1)}$ (eV)	r_{II} (Å)	$\Delta H = \dfrac{P_0\gamma}{r_{II}(n^*+1)}$ (eV)	ΔH calcula-tion $\left(\mathrm{kJ/mol}\right)$	ΔH^0 298 ref. data $\left(\mathrm{kJ/mol}\right)$
1	2	3	4	5	6	7	8	9	10	11
	Ga	$4s^2\,4p^1$	20.760	1	1.39	3.178			<284.8>	273.0
	Ga		8.8961	2	1.39	2.723				
36	In	$5s^2\,5p^1$	21.841	1	1.66	2.6314			<241.5>	238.1
		$5s^2\,5p^1$	21.841	½			0.92	2.374		
	Tl	$6s^2\,6p^1$	22.012	½			1.05	2.015	194.5	181.0
	C	$2p^2$	10.061	1	0.77	4.3554			<723.9>	716.7
		$2s^2\,2p^2$	24.585	1	0.77	10.643				
46	Si	$3p^2$	10.876	2	1.17	4.648			448.6	452
	Ge	$4p^2$	12.072	2	1.39	3.3656			<378.3>	376.6
		$4p^2$	12.072	2	1.24	4.1428				
	Ge	$4p^2$	12.072	1			0.65	3.9516	381.43	376.6
	Sn	$5p^2$	13.009	2	1.58	3.2934			317.9	302.1
	Pb	$6s^2\,6p^2$	32.526	1	1.50	2.0952			202.2	195.1
	Pb	$6p^2$	13.460				1.26	2.0543	198.3	195.1

TABLE 4 *(Continued)*

Group	Atom	Orbitals	P_0 (eVÅ)	γ	r (Å)	$\Delta H = \dfrac{P_0\gamma}{r(n^*+1)}$ (eV)	r_{II} (Å)	$\Delta H = \dfrac{P_0\gamma}{r_{II}(n^*+1)}$ (eV)	ΔH calculation (kJ/mol)	ΔH^0_{298} ref. data (kJ/mol)
	N	$2p_r{}^5$	21.966				1.48	4.947	477.5	472.7
56	P	$3p^1$	7.7864	2	$r_{\hat{e}} = 1.16$	3.3562			323.9	316.3
	As (5)	$4s^2\,4p^3$	40.232	½	1.40	3.057				295.1
	As (3)		18.645			3.2785				316.5	301.8
		$4p^3$		1	$r_{\hat{e}} = 1.21$					
	As (3, 5)								<305.8>	
1	2	3	4	5	6	7	8	9	10	11
56	Sb	$5p^3$	20.509	1		2.9509			<265.2>	268.2
		$5p^3$	20.509	1	$r_{\hat{e}} = 1.39$, 1.61	2.5477				
	Bi	$6p^3$	21.919	1	1.82	2.3160			<207.3>	209.2
	Bi	$6p^3$	21.919	1			2.13	1.9790		

TABLE 4   *(Continued)*

Group	Atom	Orbitals	P_0 (eVÅ)	γ	r (Å) Δ $H = \dfrac{P_0\gamma}{r(n^*+1)}$ (eV)	r_H (Å)	$\Delta H = \dfrac{P_0\gamma}{r_H(n^*+1)}$ (eV)	ΔH calculation $\left(\text{kJ}/\text{mol}\right)$	ΔH^0 298 ref. data $\left(\text{kJ}/\text{mol}\right)$
	O (2)	$2p^2$	11.858	1		1.40	2.823	<253.1>	249.2
66	O (4)	$2p^4$	20.338	½		1.40	2.421		
	S (2)	$3s^1\,3p^1$	21.673	2	2.882 $r\,ê =0.94$			<280.8>	277.0
	S (4)	$3p^4$	21.375	1		1.82	2.9360		
	Se (4)	$4p^4$	24.213	½	2.2032 $r\,ê =.17$			<226.7>	223.4
	Se (2)	$4s^1\,4p^1$	23.283	1		1.98	2.495		
	Te	$5s^1\,5p^1$	23.882	1		2.11	2.264	218.5	215.6
	Po	$6s^1\,6p^1$	23.664	½	1.516 $r\,ê = 1.50$			146.4	146
	F	$2p^1$	6.635	½		1.33	0.83145	80.3	79.5
76	Cl	$3p^1$	8.5461	1		1.81	1.2804	123.6	121.3
	Br	$4p^1$	9.3068	1		1.79	1.1062	106.8	111.8
	I	$5p^1$	9.9812	½		$r^{5+} = 0.94$	1.062	102.5	106.8

KEYWORDS

- **Dissociation energy**
- **Lagrangian equation**
- **Single-atom gas**
- **Spatial-energy characteristics**
- **Sublimation**

REFERENCES

1. Lebedev, Yu. A.; and Miroshnichenko, E. A.; Thermal Chemistry of Steam-Formation of Organic Substances. Ed. M. "Nauka," **2012**, 216 p.
2. Lebedev, Yu. Properties of Inorganic Compounds. Reference-Book. Efimov, A. I.; et al. Ed. L. "Khimiya," **2012**, 392 p.
3. Lebedev, Yu. Rupture Energy of Chemical Bonds. Ionization Potentials and Affinity to an Electron. Reference-Book. Kondratyev, V. I.; et al. Ed. M. "Nauka," **1974**, 351 p.
4. Benson, S.; Thermochemical Kinetics. Ed. M. Mir, **2012**, 308 p.
5. Berlin, A. A.; Wolfson, S. A.; and Enikolopyan, N. S.; Kinetics of Polymerization Processes. Ed. M. "Khimiya", **2012**, 319 p.
6. Korablev, G. A.; Spatial-Energy Principles of Complex Structures Formation. Netherlands: Brill Academic Publishers and VSP; **2005**, 426 p. (Monograph).
7. Korablev, G. A.; Spatial-Energy Criteria of Phase-Formation Processes. Izhevsk: Publishing House "Udmurt University," **2008**, 494 p.
8. Fischer, C. F.; Average-energy of configuration Hartree–Fock results for the atoms helium to radon. *At. Data.* **2012**, *4*, 301–399.
9. Waber, J. T.; and Cromer, D. T.; Orbital radii of atoms and ions. *J. Chem. Phys.* **1965**, *42(12)*, 4116–4123.
10. Clementi, E.; and Raimondi, D. L.; Atomic screening constants from S.C.F. Functions, l. *J. Chem. Phys.* **1963**, *38(11)*, 2686–2689.
11. Clementi, E.; and Raimondi, D. L.; Atomik Screening Constants from S.C.F. Functions, 2. *J. Chem. Phys.* **1967**, *47(4)*, 1300–1307.
12. Korablev, G. A.; and Zaikov, G. E.; Energy of chemical bond and spatial-energy principles of hybridization of atom orbitals. *J. Appl. Polym. Sci. USA.* **2006**, *101(3)*, 283–293.
13. Korablev, G. ; A Thermodynamic properties of individual substances. Reference-book of academy of science of the USSR, Ed. Glushko V. P.; M.: "Science," **1978**, *1,* 496 p.
14. Kofstad, P.; Deviation from stoichio, diffusion and electrical conductivity in simple metal oxides. Ed. M. "Mir," **2012**, 398 p.
15. Pound, G. M.; S. *Phys. Chem. Ref. Data.* **1972**, *1,* 135–146.

16. Korablev, G. A.; and Solovyev, S. D.; Energy of diffusion activation in metal systems. *Bull. ISTU.* **2007,** *4,* 128–132.
17. Zaharov, M. A.; Grid Models of Multi-Component Solid Solutions: Statistic Thermodynamics and Kinetics. Abstract of Doctoral Thesis, Novgorod State University, **2008,** 36 p.
18. Geguzin, Ya. E.; Ascending diffusion and diffusion aftereffect UFN. **2012,** *149(1),* 149–159.
19. Batsanov, S. S.; and Zvyagina, R. A.; Overlap Integrals and Problem of Effective Charges. "Nauka," Novosibirsk: Siberian Branch; **1966,** 386 p.
20. Batsanov, S. S.; Structural chemistry. Facts and dependencies. M.: MSU – **2012,** 292 p.

CHAPTER 7

NANOHYDRODYNAMICS AND LIQUID–SOLID INTERFACES: A COMPREHENSIVE REVIEW

A. K. HAGHI, A. POURHASHEMI, and G. E. ZAIKOV

CONTENTS

7.1   OVERVIEW

Important developments have occurred in recent years in the new field of fluid mechanics associated with nanotechnology—nanohydromechanics.

In general, nanotechnology is a set of methods for production of items with a given atomic structure by manipulating atoms and molecules. Traditional hydrodynamics is associated with motion of gases and liquids at macroscopic scale. Micro- and nanohydromechanics is the area of mechanics where one studies the motion of gas and liquids in amounts conventionally related to nanotechnology (less than 100 nm = 0.1 μm).

Internal fluid dynamics describes the gas flow in micro- and nanochannels and tubes (including nanotubes). Smallest diameter pipes in the nature are carbon nanotubes (CNTs). Medications are often delivered into the patient's body through the microholes (gramicidin ion channel has a pore diameter of 0.4 nm and length 2.5 nm).

The size and surface properties of nanoparticles that enter the body through the respiratory system affect the way they are moving throughout the body. The study first published by the US scientists may be useful for both development of sanitary standards on working with nanoparticles and new drugs development.

The extent and method of purifying lungs using nanoparticles that enter through the respiratory system depends on the size and surface characteristics of nano objects. Akira Tsuda (Harvard School of Public Health) studied how nanoparticles created by incomplete combustion processes or industrial emissions are absorbed by the body and/or removed from it. Teaming up with John Frangioni's group, Tsuda studied how these nanoparticles travel throughout the body.

The researchers studied the behavior of various fluorescent nanoparticles that differed from each other in size and composition. These nanoparticles were injected into the lungs of rats, and researchers used real-time fluorescence spectroscopy to observe how the particles are absorbed by the body, migrate through the organs of rodents, and how many such particles are excreted from the body an hour after they enter into the body.

It was found that most of the nanoparticles remain in the lungs; however, if the diameter of nanoparticles is less than 34 nm, they migrate to the lymph nodes. The speed of such movements depends on their size. Larger nanoparticles migrate more slowly. The speed of movement also depends on surface characteristics. Nanoparticles with zwitter-ionic, anionic, and polar surface migrate to the lymph nodes, while nanoparticles with cationic surface remain in the pulmonary cells. Nanoparticles with zwitter-ionic surface with the size lesser than 6 nm moved quickly into the bloodstream and then were excreted by the kidneys.

Thus, it is clear that the fluid flow through micro- and nanotubes represents a fundamental interest for many biological and technical devices and systems. Therefore, flows in the nanometer size channels are intensively studied currently.

Importance of modeling in nanohydromechanics is supported by the results of numerous experiments conducted over the past two decades that revealed significant differences in the behavior of fluids in volumes with dimensions of about tens of molecular diameters or less from the predictions of the classical continuum theories. The analysis shows that in microtubules of 50 nm in diameter, the flow is continuous; whereas in microtubules with a diameter of 5 nm it is not continuous, that is, there is strong difference in the fluid–wall interaction in the range of 5–50 nm.

Increased carrying ability of the membranes of carbon nanotubes. CNT is a channel whose diameter is several times larger than the characteristic size of atomic particles, thus CNT can be considered as a reservoir for storage of gaseous and liquid substances. Recently, the joint experiment of two groups— Lawrence Livermore Natural Laboratory and University of California, Berkeley (USA)—showed that nanotubes could serve as a channel for the transport of such substances with a capacity of 2–3 orders of magnitude higher than the corresponding quantities determined by the classical gas dynamics.

In the experiment, the film of closely packed ($\sim 2.5.10^{11}$ cm2) of vertical double-layer CNTs were grown on a silicon chip by chemical vapor precipitation in the presence of a catalyst. The space between the tubes filled with silicon nitride (Si3N4) to gas or liquid is passed only through the inner cavity of nanotubes. Excess of silicon nitride was removed from both ends of the chip by ion milling, which resulted in a nanotube with both ends opened up. Measurements made during the passage through the nanotube of colloidal gold particles of various sizes showed that these membranes are able to pass the particles with lateral dimensions between 1.3 and 2 nm.

Bandwidth ability of obtained membranes were determined for water as well as for the following gases—H_2, He, Ne, N_2, O_2, Ar, CO_2, CH_4, C_2H_6, C_3H_6, C_4H_6, and C_4H_8. Measurements were taken Xe in the Knudsen regime. The ratio of the characteristic path length of gas molecules to the diameter of nanotubes is much greater than unity and ranges from 10 to 70. The bandwidth ability of the membrane for various gases was not the same.

Similar properties of membranes prompted scientists from the Lawrence National Laboratory and University of California (USA) under the leadership Olgica Bakajin to create a quick filter of nanotubes [1,2].

Scientists demonstrated the filter element where nanotubes play the role of membrane. It was planned to use for a study of transport of gases and fluids through the CNTs with diameter less than 2 nm. Based on a previously developed mathematical model, it was assumed that the rate of passage of substances through the nanotube has to be very high, which turned out to be true.

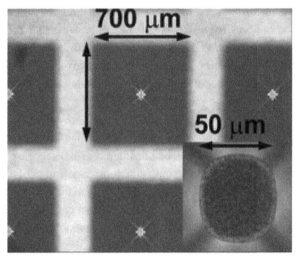

FIGURE 1   Core of a nanofilter.

The core of nanofilter is a matrix of vertically arranged double-layer nanotubes synthesized by precipitation in vapor phase on silicon nitride substrate (Figure 1). Because there was free space filled by silicon nitride and ends of nanotubes opened up by etching, researchers obtained the filter element.

Due to the high density of nanopores per unit area of the nanotube, the physics of fluid dynamics is different from the classical dynamics. As scientists proved, the filter with pore density $2.5.10^{11}/cm^2$ passes the fluid at a rate 100–100,000 times faster than predicted by the classical theory of liquids. This result allows the development of commercially successful water and gas filtration and water desalination systems. Compared with polycarbonate filters, the nanotubes are characterized by lower limit of the passing particles.

The difference in the speed of passage of gas through the membrane of different masses of the CNT indicates the selectivity of the migration process. This can be used for solving tasks of the gas separation of different sorts or different isotopic modifications. Multiple excess overlook of membranes based on CNTs over the value characteristic of the Knudsen regime is due to the changing nature of the interaction of gas molecules with the inner walls of the nanotubes compared to the macroscopic surface.

The inner surface of the nanotubes is smooth on length scales down to atomic scales. At the same time, the macroscopic surfaces of porous materials have a roughness on much larger scale. For this reason, the nature of the interaction of atomic particles with the walls of the nanotubes largely corresponds to mirror reflection and not diffuse reflection, as is in the case of macroscopic surfaces. Thus, the gas inside CNT feels much less resistance from the surface than is determined by classical expression for the Knudsen flow.

Membranes based on CNTs are able to pass not only gases but also liquid substances. This experiment revealed that the overlook properties of membranes for water are more than three orders of magnitude higher than the corresponding value calculated on the basis of the classical formula of Hagen–Poiseuille. This effect is also related to the difference of the interaction of liquid with the inner walls of CNTs compared to the macroscopic surface. *Features of micro- and nanohydromechanics:* a very large ratio of surface to volume measurements; comparable sizes of channel itself and molecules moving through the channel; possibility of significant density fluctuations in contrast to makroflows; and transport properties (viscosity, diffusion, thermal conductivity) might contain the size factor (as in turbulence); nanoflow interaction with the wall might be the determining factor; there is no exact boundary conditions; a continuum approximation can be violated; some events observed in micro- and nanoflows just do not exist in macrohydrodynamics.

At the nanometer scale liquid shows unusual properties, for example, an abrupt increase in viscosity and density near the walls of nanocapillaries, the change of thermodynamic parameters of the liquid, as well as atypical chemical activity at the border between solid and liquid phases. The experiments discovered significant increase in the effective viscosity of the fluid compared to its macroscopic value. This viscosity depends on the nanotube diameter.

The effective viscosity of the liquid in a nanotube is determined as follows: Let us compare two nanotubes of the same size and under the same pressure—one filled with liquid possibly containing crystallites and another filled with liquid considered a homogeneous medium (i.e., without considering the crystallite structure) where Poiseuille flow is realized. The viscosity of a homogeneous fluid that ensures the same mass transport for both tubes is the effective viscosity of the flow in the first nanotube.

Channel width reduction leads to the increase of Knudsen number and a rise in the role of surface interactions. And macroscopic consideration of natural gas as a continuous medium becomes invalid. Therefore, it is accepted to use microscopic approach based on the methods of kinetic theory and direct statistical and molecular dynamic simulations.

Classical hydrodynamics, not taking into account atomic (molecular) structure of the liquid, does not adequately describe the flow of liquids in nanochannels with a width of about 10 molecular diameter or less. It is a major problem for nanotechnology researchers—the laws of classical physics no longer apply.

The next question is related to the use of classical or quantum hydrodynamics for describing nanoflow. Quantum hydrodynamics is important where the laws of classical physics become broken. And it is clear that due to much smaller scale the number of such problems in the micro- and nanohydromechanics is significantly bigger than in classical hydrodynamics, in particular, in a heat transfer from fluid to the wall (phonons) and phenomena that involves electrons (electromagnetic phenomena).

As a rule, the factors affecting nanotechnology applications are complex, that is, electro-, hydraulic-, magnetic, optical, and other processes are all important. To solve quantum hydrodynamics problems, one would take into account the processes where quantum effects are important.

The practical use of nanotubes. One of the most attractive destinations for using nanotubes is microelectronics. Small size, possibility to obtain the needed electrical conductivity during synthesis, mechanical strength, and chemical stability make nanotubes very desirable materials for microelectronics.

Nanotubes have many interesting properties. For example, two nanotubes of the same chemical composition might have different conductivity depending on their configurations of atoms in the "lattice" (one to be closer to "metallic" and the other to "solid state").

Theoretical calculations have shown that if the defect in the form of a pentagon–heptagon pair is embedded into an ideal single-walled nanotube with chirality (8, 0), the chirality of the tube in the vicinity of the defect becomes (7, 1). A nanotube with chirality (8, 0) is a semiconductor with a band gap of 1.2 eV, whereas the nanotube with chirality (7, 1) is a semimetal with zero band gap. Thus, a nanotube with an embedded defect can be considered a hetero-conducting super small metal-semiconductor element.

Currently, the efforts of scientists are focused on developing technologies to obtain CNTs filled in with conducting or superconducting materials. The goal is to create conductive compounds—the base for production of the smallest nanoelectronic devices.

Individual nanotubes can be used as delicate probes for the study of surfaces with roughness on the nanometer level. In this case, the extremely high mechanical strength of the nanotube is utilized. The modulus of elasticity E along the longitudinal axis of the nanotube is about 7,000 GPa, whereas the probes made of steel and iridium barely reach the values of E = 200 and 520 GPa, respectively. In addition, single-walled nanotubes, for example, may get elastically lengthened by 16 percent. To visualize such a material property from 30 cm iron spoke, it should be lengthened by 4.5 cm under load and then get back to the original length. The probe of the nanotube with superelastic properties will bend elastically with the load exceeding certain level, thereby providing contact with the surface.

High modulus of elasticity of CNTs allows creation of composite materials having high strength during ultra-high elastic deformation. And it can be successfully used to produce ultra-light and heavy-duty fabric for clothing of firefighters and astronauts.

The strength of a material is its ability to withstand an applied stress without failure—either fracture (separation of the parts) or irreversible change in shape (plastic deformation). When a cylindrical sample with a cross section of S gets stretched by force F, it gets deformed elastically at first (reversible deformation, see the section O on the curve in Figure 2), and then plastically, that is, irreversible (see section Π in Figure 2).

At the top of Figure 2, you can see a schematic drawing of a red cylindrical sample with a cross-sectional area S under applied force F getting extended by an amount of $L - L_0$, where L_0 is the original sample length. At the bottom, there is a drawing of relationship between mechanical pressure and relative deformation with arrows indicating the results for titanium, steel, and bronze.

While sample gets deformed, its structural inhomogeneities (lattice defects or dislocations) start moving, colliding with others, thus forming microcracks. The more the dislocations exist in the sample and the sooner they start moving, the more the micorcracks formation.

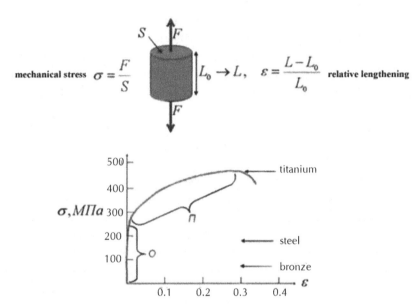

FIGURE 2   Schematic illustration of red cylindrical sample and the relationship between mechanical stress and relative lengthening during tension specimen.

When σ ($\sigma = F/S$) reaches a certain level, then the neighboring microcracks connect to each other reaching critical size and the sample gets destroyed.

Nanowire is a single crystal virtually free from defects (dislocations). In addition, the surface of nanowire has very small radius of curvature (10 nm); it is highly compressed, and therefore prevents the movement of dislocations out, that is, formation of microcracks. All these lead to the fact that nanowire has almost no plastic deformation and its tensile strength is 10 times higher than for normal samples (see Figure 3).

Figure 3 shows the relationships between stress and relative deformation during tensile tests of microsamples of various diameters made from Ni and its alloys $Ni_3Al - Ta$. [3] The diameter values are shown next to corresponding curves. For comparison, the red shows the "stress-relative strain" for macro samples. It shows that the tensile strength increases with decreasing diameter of wire.

Let us calculate the strength of CNT using single-wall of "zig-zag" nanotube (Figure 4). We lock the invisible end of the tube and apply tensile force F to the other end. Figure 4 shows blue lines crossing C–C links oriented along the axis of the tube, and the yellow arrow shows the direction of the tensile force F .

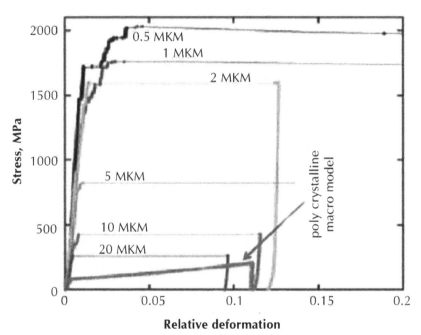

FIGURE 3 The relationship between stress and relative deformation during tensile microsamples of different diameters of Ni and its alloys $Ni_3Al - Ta$.

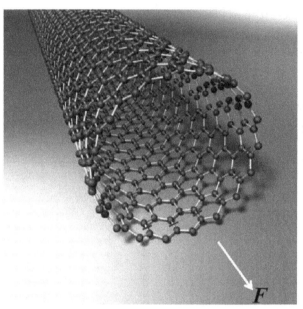

FIGURE 4   Carbon nanotube schema.

Let carbon atoms form identical connections (C–C, σ-connections) in the nanotube and the angles between them be equal to 120°. Then, when nanotube gets stretched, these connections will stretch in the same manner. And nanotube can explode in a most unexpected way depending, for instance, which C–C connection breaks first.

To simplify the calculations, let us assume that the tension breaks only C–C connections oriented along the axis of the tube and located on the same cross-section (blue break lines on Figure 4).

It is known that the distance between adjacent carbon atoms in the nanotube is approximately equal to $d = 0.15$ nm. It is easy to show that if the tube diameter is D, then the amount of N connections oriented along the axis of the tube is

$$N = \frac{\pi D}{\sqrt{3}d} \tag{1}$$

Value of force applied to each C–C connections equals F/N.

The strength of C–C connection can be obtained from a curve in Figure 5 that shows dependence of connection potential energy on the distance between atoms.

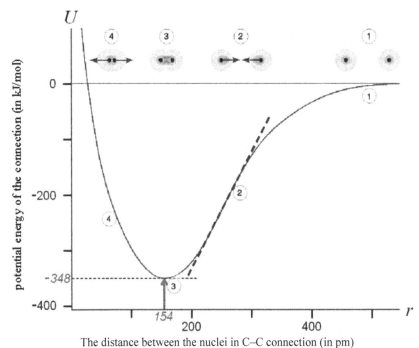

FIGURE 5   How potential energy of C–C connection mole (i.e., 6.10^{23}) depends on the distance between the nuclei.

Figure 5 diagram shows that the potential energy of connection reaches the minimum when the distance between the nuclei of atoms is 154 pm. This determines the distance between carbon atoms when nanotube is not stretched.

Tangent of slope of the right branch of the curve in Figure 5 is proportional to the force F_1 needed to hold the atoms at a given distance r :

$$F_1 = \frac{\partial U}{\partial r} \cdot \frac{1}{N_A},$$

where N_A—Avogadro's number, $6 \cdot 10^{23} \, \mathrm{mol}^{-1}$.

To increase the distance between the carbon atoms, one has to apply force F_1, and if it is greater than the maximum tangent of the slope (see blue dotted line in Figure 5), then C–C connection breaks. It happens when

$$F_1 > \frac{348 \cdot 10^3}{154 \cdot 10^{-12} \cdot 6 \cdot 10^{23}} = 3{,}8 \; \textit{нН} \tag{2}$$

The nanotube will break when the force F stretching the tube becomes more than $3.8N$ nN, where N is the number of C–C connections parallel to the axis of a single cross section of the tube. If nanotube diameter $D = 1.5$ nm, then $N = 18$ (from (1)). Therefore, the nanotube will break at $F_{max} > 69$ nN.

To calculate the strength σ_{max} of the nanotube, let us divide F_{max} by the cross-sectional area number $S = \pi D^2 / 4$:

$$\sigma_{max} = \frac{F_{max}}{\pi D^2} \cdot 4 = \frac{4 \cdot 69 \cdot 10^{-9}}{3{,}14 \cdot 2{,}25 \cdot 10^{-18}} = 39 \; \textit{ГПа} \tag{3}$$

This value is quite close to the experimentally obtained values maximum (63 GPa) and, as expected, much more than the strength of the most strong types of steel (0.8 GPa).

Note that multiwalled nanotube strength is several times larger!

Another attractive nanotube property is its very high surface area. In the process of growth, randomly oriented helical nanotubes are formed, which leads to a significant number of cavities and voids of nanometer size. As a result, the area per unit mass of nanotube material reaches values of about 600 m^2/g. Such a high surface area opens up the possibility of usage in filters and other devices of chemical technology.

The unique chemical properties of CNTs allow very small molecules of water to seep through, whereas viruses, bacteria, toxic metal ions, and toxic organic molecules cannot do this.

Thanks to the amazing properties, nanotubes can be used in various technological fields, ranging from the creation of composite materials and heavy-duty thread, to the construction of nanodevices and nanosensors.

Nanotubes can serve as the storage for hydrogen, the cleanest fuel.

The reserves of coal, oil, and gas on the Earth are limited. In addition, the burning of conventional fuels leads to the accumulation of carbon dioxide and other harmful pollutants in the atmosphere, and this in turn leads to global warming, the signs of which humanity has already experienced. Thus, today humankind is facing a very important problem—to replace traditional fuels in the not so distant future.

It is really beneficial to use hydrogen—the most common chemical element in the universe—as a fuel. The oxidation (combustion) of hydrogen proceeds with the release of a very large amount of heat (120 kJ/kg) in addition to water. For comparison, the combustion of gasoline or natural gas releases three times less heat than that of hydrogen. It should also be noted that the combustion of hydrogen does not have any harmful effects on the environment.

There are quite a few fairly cheap and environmentally friendly ways to produce hydrogen; however, hydrogen storage and transportation has so far been an unresolved problem. The reason for this is the very small size of hydrogen molecule. Hence, hydrogen can leak through microscopic cracks and pores of conventional materials, and it might cause explosion. Thus, the walls of containers for storing hydrogen should be thicker and therefore heavier. For more safety, it is better to freeze hydrogen to several dozen $^\circ K$, which raises the price of storing even further.

The solution to the above-mentioned problem could be a device that plays the role of "sponge"—with the ability to absorb hydrogen and hold it indefinitely. Obviously, such a hydrogen "sponge" must have a large surface area and chemical affinity to hydrogen. All these properties are present in CNTs.

It is well known all the atoms of CNT are on its surface. It is the base for one of the main mechanisms of how nanotube absorbs hydrogen—chemisorption, that is, the adsorption of hydrogen H_2 on the surface of the tube with subsequent dissociation and formation of chemical connections of C–H. Absorbed in such a way, hydrogen can be extracted from nanotubes, for example, by heating it to 600°C. In addition, the hydrogen molecules are bound to the surface of the nanotubes due to physical adsorption by van der Waals interactions.

Researchers figured out that any efficient fuel cell based on hydrogen "sponge" should contain at least 63 kg of hydrogen per cubic meter. In other words, the weight of stored hydrogen must be not less than 6.5% of the weight of the "sponge." Currently, the experimental "sponge" can store more than 18 percent of hydrogen, and it opens up broad prospects for development of hydrogen energy.

Suetin, M. V., and Vakhrushev, A. V. [4], investigated nanocapsule for storing methane using methods of molecular dynamics. Nanocapsule consisted of three nanotubes: (20,10) (10,10), and (8.8) joined together.

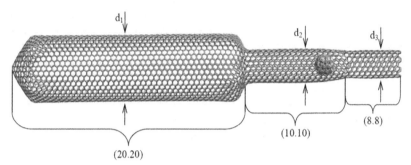

FIGURE 6 Nanocapsule for natural gas storage, consisting of three types of nanotubes (20,20), (10,10), (8,8), and containing $K@C_{60}$.

There was the endohedral complex $K@C_{60}$ with single positive charge inside nanocapsule that was used for blocking. Its movement was controlled by external electric field.

The modeling was done by molecular dynamics methods with 1fs step. The object of the study was nanocapsule whose structure is shown in Figure 6.

The nanocapsule consists of "armchair"-type nanotubes of different diameters: (20,20)—$d_1 = 25.93A = 2.593$ nano ($1Å = 10^{-10} M = 0.1$ H), (10,10)—$d_2 = 13.4A$ and (8,8)—$d_3 = 10.74A$, combined together by heptagonal rings ($d = 0.1295 \cdot (n,n)$). Nanocapsule contains endohedral complex $K@C_{60}$ where potassium atom has a single positive charge. The @ symbol means that K is contained inside C_{60}. The diameter d_2 of nanocapsule at the site (10,10) is big enough to let $K@C_{60}$ get in; however, note that the nanocapsule is big enough to allow gas molecules to move between $K@C_{60}$ and the wall of nanocapsule. The diameter d_3 of nanocapsule at the site (8,8) is not sufficient for $K@C_{60}$ to pass, but good enough for methane molecules. Region of nanocapsules (20,20) is used to accumulate and store gas molecules. Endohedral complex $K@C_{60}$ in nanocapsule is moved by external electric field.

Nanocapsule works in three modes: adsorption, desorption, and storage. At the stage of the adsorption, $K@C_{60}$ is located near the base of nanocapsule (left end of the nanocapsule—Figure 6). Methane molecules pass through the hole in the nanocapsule and get adsorbed on the walls. To switch to the storage mode, it is necessary to seal the nanocapsule entrance. $K@C_{60}$ moves by electric field to block the entrance—now methane molecules cannot leave the nanocapsule. It stops at (8,8) site unable to move any further and stays there even after disappearance of the electric field—due to the pressure of methane molecules and capillary forces. At this stage (storage), the external thermodynamic conditions return to normal.

To initiate the stage of desorption, $K@C_{60}$ is pushed to the bottom of the nanocapsule. Due to excess pressure inside the nanocapsule, methane molecule starts leaving its inner space. However, a significant part of methane remains in the nanocapsule concentrated in the areas of nanotubes (10.10) and (8.8). For complete extraction of methane, $K@C_{60}$ is moved to the site (10,10) and squeezes the methane out. Then $K@C_{60}$ is again pushed to the bottom of the nanocapsule. Methane molecules are concentrated in the area of (10,10) again, and then $K@C_{60}$ squeezes them out. This happens as long as all methane molecules leave the interior of the nanocapsule.

The nanocapsule shown in Figure 6 contains 737 molecules of methane. The pressure inside the nanocapsule is about 10 MPa. The density of methane inside nanocapsules is ~82 kg/m³. The weight content is calculated as follows:

$$W_t = \frac{N_{CH_4} \times m_{CH_4}}{N_{CH_4} \times m_{CH_4} + N_C \times m_C} \times 100\%$$

,

where N_{CH_4}—number of methane molecules, N_C—number of carbon atoms in the nanotube, m_{CH_4}—mass of one molecule of methane, and m_C—mass of one carbon atom. Thus, there is ~17.5 wt % of methane in the nanocapsule.

Figure 7 shows two graphs for desorption stage: the first is $K@C_{60}$ velocity V and the second is the $K@C_{60}$ coordinate d. $K@C_{60}$ moves under the influence of an electric field from the site of nanotube (10,10) to the center of nanocapsule base area of nanotube (20,20).

Strength of the electric field ($5.14 \cdot 10^9$ V/m) is determined by the need to overcome the capillary forces that oppose $K@C_{60}$ move from the site (10,10) to (20,20) plus the pressure of compressed methane.

Nanocapsule is similar to bottle-like pore, hence a significant amount of methane molecules in the pore remains there even after the nanocapsule is opened. In our case, about 7.9 wt % of methane remains in the nanocapsule. Given that the adsorption potential at the sites (10,10) and (8,8) is significantly higher than at (20,20), the methane molecules are concentrated in the first two sections. Now, it is necessary to desorb the gas molecules. To do this, $K@C_{60}$ starts moving through electric field (magnitude $5.14 \cdot 10^9$ v/m is needed) toward the (10,10) site of the nanotube and squeezes methane from the nanocapsule.

To do this, $K@C_{60}$ moves by electric field (magnitude $5.14 \cdot 10^9$ v/m is needed) toward the (10,10) site of the nanotube and squeezes methane from the nanocapsule.

Figure 8 shows the curve of methane molecules desorption from nanocapsule. As one can see, it is almost linear while $K@C_{60}$ is moving from the site (10,10) and the desorption ends abruptly with $K@C_{60}$ reaching the site (8,8).

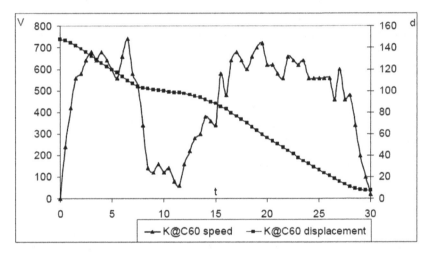

FIGURE 7   How $K@C_{60}$ velocity V and $K@C_{60}$ coordinate d behave during nanocapsule desorption.

To get $K@C_{60}$ back, the direction of the electric field should be reversed (with the same magnitude). Hence, a single cycle of desorption removes about 1.4 wt % of methane from the nanocapsule.

Meaning that if nanocapsule contains 7.9 wt % of methane then the complete desorption is achieved by six cycles of $K@C_{60}$ extrusion. Of course, we should keep in mind that certain amount of methane might still remain in the nanocapsule, but higher adsorption potential of (10,10) area leads us to believe that the vast majority of methane molecules is concentrated precisely at (10,10) area.

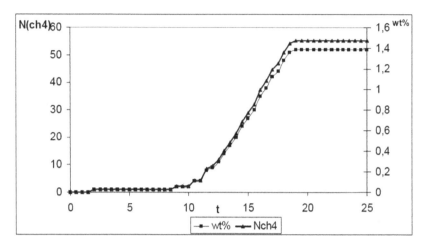

FIGURE 8   Quantity (N (ch4)) and weight (wt %) of methane molecules in the nanocapsule during $K @ C_{60}$ extrusion.

As one can see, the nanocapsule is fairly complex structure where the locking element $K @ C_{60}$ behavior is controlled by an external electric field. And its direction and magnitude of tension determine the position of $K @ C_{60}$ in nanocapsule and accordingly the phases of adsorption, desorption, and storage.

Here are some possible applications of nanohydromechanics.

Microbubble medium (gas in liquid). This is an interesting area with multiple applications. And any success in the area requires significant efforts in several departments: studying of thermodynamics of two-phase systems with free boundary, as applied to micro- and nanobubble environments; estimates of energy costs needed to obtain microbubble environments (there are various ways to do it); the kinetics of growth and destruction of microbubbles in the liquid; limits of tension leading to the destruction of microbubbles; physics of simple liquids, as applied to microbubbles mediums; development of mathematical models to describe the physical properties of micron- and nanosized bubbles of media; theoretical methods and numerical modeling of viscosity, density, and sedimentation stability of micro- and nanobubble media; analysis of opportunities for modifying the properties of the liquid in hydrodynamic devices; carrying out theoretical studies of possible ways to get microbubbles nanoscale environments; atomic force microscopy in nanobubble environments; nanobubbles in sonoluminescence.

Microhydromechanics of oil. Behavior of oil in reservoirs at a depth of 1–3 km in hard rocks is determined using Darcy law. Typical values of permeability range from 5 to 500 mD. The permeability of coarse-grained sandstone is $10^{-8} - 10^{-9}$ sm^2 and the permeability of tight sandstone is around 10^{-10} sm^2. For reservoirs with permeability as a fraction of micron, the flow of oil is described by nanotechnology methods.

Microstructure of a viscoplastic fluid. In many cases, the liquid changes its properties due to irreversible processes occurring at micro- and/or nanoscale; thus, there is a change in the rheological properties of liquids. Therefore, both experimental and theoretical studies of nanorheology gels used in hydrofracking are becoming more relevant.

Micro- and nanohydromechanics is a new area of fundamental and applied mechanics where fundamental research is the basis for creation of real-world hydrodynamic nanotechnologies—the core of several promising twenty-first century technologies.

The main method of nanotubes manufacturing is arc discharge process. The anode "evaporates" and multilayered nanotubes "grow" on the surface of the cathode.

Until now, nanotubes obtained for a variety of studies have not been able to provide the required purity of the experiment. The tubes have always contained some dirt-making precise measurements impossible: the electron can be influenced by an additional electric potential caused by pollution making the results meaningless.

7.1.1 STRUCTURES OF A CRYSTAL LATTICE OF DIAMOND AND GRAPHITE

In 1991, Idzhima studied the sediments formed at the cathode during the spray of graphite in an electric arc. His attention was focused on the unusual structure of the sediment consisting of microscopic fibers and filaments. Measurements made with an electron microscope showed that the diameter of these filaments does not exceed a few nanometers and a length of one to several microns.

Having managed to cut a thin tube along the longitudinal axis, the researchers found that it consists of one or more layers, each representing a hexagonal grid of graphite, which is based on hexagon with vertices located at the corners of the carbon atoms. In all cases, the distance between the layers is equal to 0.34 nm, which is the same as that between the layers in crystalline graphite.

Typically, the upper ends of tubes are closed by multilayer hemispherical caps; each layer is composed of hexagons and pentagons, reminiscent of the structure of half a fullerene molecule.

The extended structure consisting of rolled hexagonal grids with carbon atoms at the nodes are called nanotubes.

Lattice structure of diamond and graphite are shown in Figure 9. Graphite crystals are built of planes parallel to each other, in which carbon atoms are arranged at the corners of regular hexagons. The distance between adjacent carbon atoms (each side of the hexagon $d_0 = 0{,}141$ nm) between adjacent planes is 0.335 nm.

Each intermediate plane is shifted somewhat toward the neighboring planes, as shown in Figure 9.

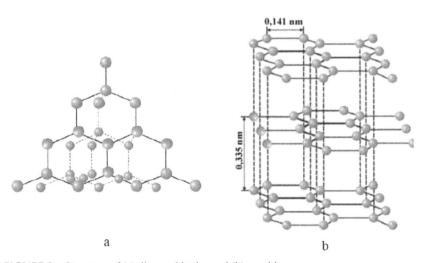

a   b

FIGURE 9   Structure of (a) diamond lattice and (b) graphite.

FIGURE 10   Schematic illustration of the graphene.

The elementary cell of the diamond crystal represents a tetrahedron, with carbon atoms in its center and four vertices. Atoms located at the vertices of a tetrahedron form a center of the new tetrahedron and, thus, are also surrounded by four atoms each. All the carbon atoms in the crystal lattice are located at equal distance (0.154 nm) from each other.

Nanotubes are rolled into a cylinder (hollow tube) graphite plane, which is lined with regular hexagons (with carbon atoms at the vertices of a diameter of several nanometers) (refer to Figure 10). Nanotubes can consist of one layer of atoms, single-wall nanotubes (SWNTs), and represent a number of "nested" one into another layer pipes, multi-walled nanotubes (MWNTs).

Nanostructures can be built from the molecular blocks as well. Such blocks or elements are graphene, CNTs, and fullerenes.

7.1.2   GRAPHENE

Graphene is a single flat sheet, consisting of carbon an atom linked together and forming a grid (each cell is like a bee's honeycombs) (refer to Figure 10). The distance between adjacent carbon atoms in graphene is about 0.14 nm.

Graphite, from which slates of usual pencils are made, is a pile of graphene sheets (Figure 11). Graphenes in graphite are very poorly connected and can slide relative to each other. Hence, if you conduct the graphite on paper, then after separating graphene from the paper the graphite remains. This explains why graphite is used for writing on paper.

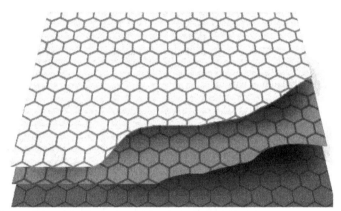

FIGURE 11   Schematic illustration of the three sheets of graphene.

The way of folding nanotubes—the angle between the directions of nanotube (Figure 12) axis relative to the axis of symmetry of graphene (the folding angle)—largely determines its properties.

FIGURE 12   Carbon nanotubes.

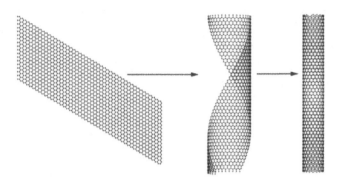

FIGURE 13   Formation of nanotube.

7.1.3   CARBON NANOTUBES

Many perspective directions in nanotechnology are associated with CNTs.

CNTs are a carcass structure or a giant molecule consisting only of carbon atoms.

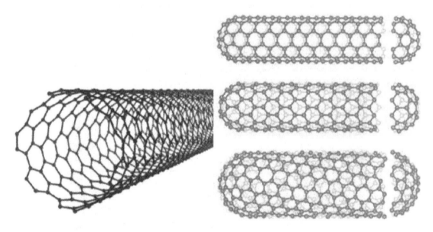

FIGURE 14   Left—schematic representation of CNTs.

CNT is easy to visualize , if we imagine that we fold up one of the molecular layers of graphite—graphene (Figure 13).

The angle between the direction of nanotube axis relative to the axis of symmetry of graphene (the folding angle) determines its properties.

Nanotubes formed by themselves, on the surface of carbon electrodes during arc discharge between them. At discharge, the carbon atoms evaporate from the surface, and connect with each other to form nanotubes (Figure 14).

The diameter of nanotubes is usually about 1 nm and their length is a thousand times more, amounting to about 40 μm. They grow on the cathode in perpendicular direction to surface of the butt. The so-called self-assembly of CNTs from carbon atoms occur. Depending on the angle of folding of the nanotube they can have conductivity as high as that of metals, and they can have properties of semiconductors.

CNTs are stronger than graphite, although made of the same carbon atoms, because carbon atoms in graphite are located in the sheets. It is known that a sheet of paper folded into a tube is much more difficult to bend and break than a regular sheet. That is why CNTs are strong. Nanotubes can be used as very strong microscopic rods and filaments, as Young's modulus of single-walled nanotube reaches values of the order of 1–5 TPa, which is much more than steel! Therefore, a thread made of nanotubes, which has the thickness of a human hair is capable to hold down hundreds of kilos of cargo.

It is true that at present the maximum length of nanotubes is usually about a hundred microns (which is certainly too small for everyday use). However, the length of the nanotubes obtained in the laboratory is gradually increasing.

7.1.4 FULLERENES

The carbon atoms, evaporated from a heated graphite surface, connecting with each other, can form not only nanotube but also other molecules that are closed convex polyhedra. In these molecules, the carbon atoms are located at the vertices of regular hexagons and pentagons, which make up the surface of a sphere or ellipsoid.

All of these molecular compounds of carbon atoms are called fullerenes on behalf of the American engineer, designer, and architect R. Buckminster Fuller, who used pentagons and hexagons for construction of domes of his buildings (Figure 15).

The molecules of the symmetrical and the most studied fullerene consists of 60 carbon atoms (C_{60}) (shown in Figure 16). The diameter of the fullerene C_{60} is about 1 nm.

FIGURE 15   Biosphere of Fuller (Montreal, Canada).

The image of the fullerene C_{60} many consider as a symbol of nanotechnology (Figure 17).

FIGURE 16   Schematic representation of the fullerene C_{60}.

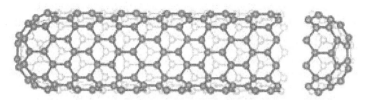

FIGURE 17   Graphical representation of single-walled nanotube.

7.1.5 CLASSIFICATION OF NANOTUBES

The main classification of nanotubes is conducted by the number of constituent layers.

Single-walled nanotubes—The structure of the nanotubes can be represented as a "wrap" hexagonal network of graphite (graphene). This is based on hexagon with vertices located at the corners of the carbon atoms in a seamless cylinder. The distance d between adjacent carbon atoms in the nanotube is approximately equal to $d = 0, 15$ nm.

Multiwalled nanotubes (Figure 18) consist of several layers of graphene stacked in the shape of the tube. The distance between the layers is equal to 0.34 nm, that is, the same as that between the layers in crystalline graphite.

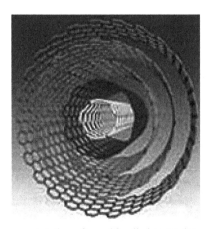

FIGURE 18   Graphic representation of a multiwalled nanotube.

Due to its unique properties (high fastness (63 GPa), superconductivity, capillary, optical, magnetic properties, etc.), CNTs could find applications in the following various applications:

- Additives in polymers
- Catalysts
- Absorption and screening of electromagnetic waves
- Transformation of energy
- Anodes in lithium batteries
- Keeping of hydrogen
- Composites (filler or coating)
- Nanosondes
- Sensors
- Strengthening of composites
- Supercapacitors

7.1.6 *CHIRALITY*

Chirality is a set of two integer positive indices (n, m), which determines how the graphite plane folds.

Values of parameters (n, m) are distinguished as follows:
- direct (achiral) high-symmetry CNTs
  - o armchair $n = m$
  - o zigzag $m = 0$ or $n = 0$
- helical (chiral) nanotube

Figure 19 shows schematic representation of the atomic structure of graphite plane.

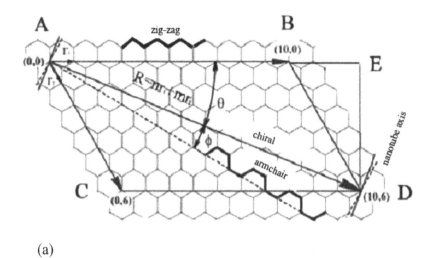

(a)

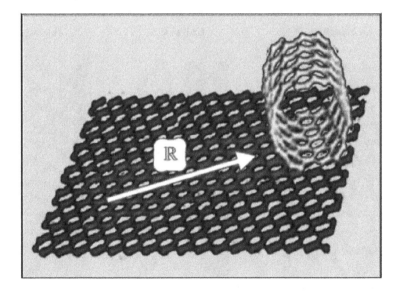

(b)

FIGURE 19   Schematic representation of the atomic structure of graphite plane.

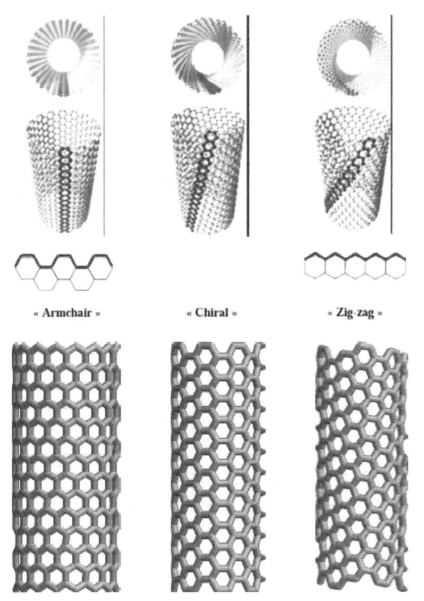

FIGURE 20   Single-walled carbon nanotubes of different chirality (in the direction of convolution).

Left to right: zigzag (16,0), armchair (8,8), and chiral (10,6) CNTs.

The cylinder is obtained by folding this sheet. To obtain a CNT from a graphene sheet, it should turn so that the lattice vector $\bar{R}$ has a circumference of the nanotube (Figure 19(b)). This vector can be expressed in terms of the basis vectors of the elementary cell graphene sheet $\vec{R} = n\vec{r_1} + m\vec{r_2}$. Vector $\vec{R}$, which is often referred to simply by a pair of indices (n, m), called the chiral vector. It is assumed that $n > m$. Each pair of numbers (n, m) represents the possible structure of the nanotube.

In other words, the chirality of the nanotubes (n, m) indicates the coordinates of the hexagon, which as a result of folding the plane has to be coincide with a hexagon, located at the beginning of coordinates (Figure 20).

Many of the properties of nanotubes depend on the value of the chiral vector. For example, a nanotube (10,10) in the elementary cell contains 40 atoms and is a type of metal. Nanotube (10,9) in 1084 is a semiconductor (Figure 21).

If the difference $n - m$ is divisible by 3, then these CNTs have metallic properties. Semimetals are all achiral tubes such as "chair." In other cases, the CNTs show semiconducting properties. The chair CNTs $(n = m)$ are strictly metal.

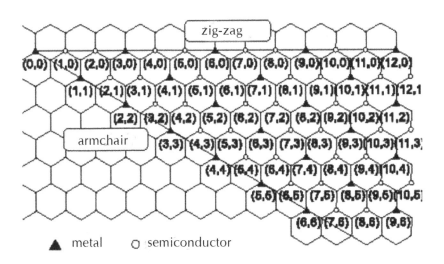

FIGURE 21 The scheme of indices (n, m) of lattice vector $\bar{R}$ tubes having semiconductor and metallic properties.

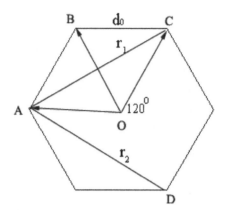

FIGURE 22   Elementary cell.

7.1.7   DIAMETER, CHIRALITY ANGLE, AND MASS OF SINGLE-WALLED NANOTUBE

Indices (n, m) of single-walled nanotube chirality unambiguously determine its diameter. Therefore, the nanotubes are typically characterized by a diameter and chirality angle. Chiral angle of nanotubes is the angle between the axis of the tube and the most densely packed rows of atoms. From geometrical considerations, it is easy to deduce relations for the chiral angle and diameter of the nanotube. The angle between the basis vectors of the elementary cell (Figure 22) r_1 and r_2 is equal to 60°.

From trigonometry, $AC^2 = OA^2 + OC^2 - 2OA \cdot OC \cdot \cos 120^0$. As $OA = OC = d_0$, $r_1 = r_2 = AC$,

$$r_1 = r_2 = \sqrt{3} \cdot d_0, \tag{1.1}$$

where $d_0 = 1,41 \overset{0}{A} = 0,141\,\text{HM}$, the distance between neighboring carbon atoms in the graphite plane.

Now, consider the parallelogram $ABDC$ in Figure 19(a).
According to Eq (1.1)

$$AB = CD = \sqrt{3}d_0 n, \quad AC = BD = \sqrt{3}d_0 m \tag{1.2}$$

Angle $\angle CAB = 60^\circ$ and $\angle ABD = 120^\circ$; therefore,

$$R^2 = 3n^2 d_0^2 + 3m^2 d_0^2 - 2 \cdot 3mnd_0^2 \cos 120^0 \text{ ,_from which we obtain}$$

$$R = \sqrt{3} d_0 \sqrt{n^2 + m^2 + mn}$$

Taking into account that $R = \pi \cdot d$, the diameter of the nanotube is determined by

$$d = \frac{\left| \overrightarrow{R} \right|}{\pi} = \sqrt{3\left(m^2 + n^2 + mn\right)} \cdot \frac{d_0}{\pi} \tag{1.3}$$

when $m = n$

$$d = \frac{3n d_0}{\pi}$$

Tables 1 and 2 summarize the values of the diameters of nanotubes of different chirality.

TABLE 1   Diameters of nanotubes of different chirality

(n, m)	d, nm	(n, m)	d, nm
(3,2)	0.334	(10,8)	1.232
(4,2)	0.417	(10,9)	1.298
(4,3)	0.480	(11,3)	1.007
(5,0)	0.394	(11,6)	1.177
(5,1)	0.439	(11,10)	1.434
(5,3)	0.552	(12,8)	1.375
(6,1)	0.517	(14,13)	1.844
(7,3)	0.701	(20,19)	2.663
(9,2)	0.801	(21,19)	2.732
(9,8)	1.161	(40,38)	5.326

TABLE 2   Diameters of nanotubes of different chirality

CNT (*n, m*)	Diameter CNT (nm)	Chirality
(4,0)	0.33	
(5,0)	0.39	
(6,0)	0.47	
(7,0)	0.55	
(8,0)	0.63	Zigzag
(9,0)	0.70	
(10,0)	0.78	
(11,0)	0.86	
(12,0)	0.93	
(3,3)	0.40	
(4,4)	0.56	
(5,5)	0.69	
(6,6)	0.81	Armchair
(7,7)	0.96	
(8,8)	1.10	
(4,1)	0.39	
(4,2)	0.43	
(7,1)	0.57	
(6,3)	0.62	Chiral
(9,1)	0.75	
(10,1)	0.82	
(6,7)	0.90	

Knowing d, the chirality m and n can be found (Tables 2 and 3). The minimal diameter of the tube is close to 0.4 nm, which corresponds to the chirality (3, 3), (5, 0), (4, 2).

We derive a formula to determine the mass of the nanotube with diameter d and length L.

The area of the elementary area; consider a parallelogram with vertices at the centers of 4 neighboring hexagons (Figure 23) with base $\sqrt{3}d_0$ and height $3d_0/2$, then $S_{re} = \dfrac{3\sqrt{3}}{2}d_0^2$.

FIGURE 23 Elementary area of graphene.

The total area of the nanotube is πdL. Consequently, the number of elementary areas is equal to $\pi dL / S_{re}$.

The mass of a CNT is equal to

$$m_T = 2m_C \frac{\pi Ld}{S_{n\jmath}} = \frac{4\sqrt{3}\pi \cdot dL}{9d_0^2} m_C \qquad (1.4)$$

where $m_C = 12$—mass of carbon atoms.

To determine the chiral angle θ from a right triangle AED

$$\sin\theta = \frac{DE}{R}, \quad \cos\theta = \frac{AE}{R} = \frac{\sqrt{3}nd_0 + BE}{R}$$

If we take into consideration $\angle EDB = 30°$, we see that $DE = \dfrac{3}{2}md_0$ and $BE = \dfrac{\sqrt{3}}{2}md_0$, consequently,

$$\sin\theta = \frac{3md_0}{2R}, \quad \cos\theta = \frac{\sqrt{3}d_0(n+m/2)}{R}$$

From these equalities, we obtain the relation between the chiral indices (m, n) and angle θ:

$$\theta = arctg\left(\frac{\sqrt{3}m}{2n+m}\right) \qquad (1.5)$$

when $m = n$

$$\theta = arctg\frac{\sqrt{3}}{3} \text{ or } \theta = 30^{0}.$$

7.2   MICRO- AND NANOSCALE SYSTEMS

Viscous forces in the fluid can lead to large dispersion flow along the axis of motion. They have a significant impact, both on the scale of individual molecules and the scale of microflows; near the borders of the liquid–solid (beyond a few molecular layers), during the motion on complex and hetero-geneous borders.

Influence of the effect of boundary regions on the particles and fluxes have been observed experimentally in the range of molecular thicknesses up to hundreds of nanometers. If the surface has a superhydrophobic property, this range can extend to the micron thickness. *Molecular theory can predict the effect of hydrophobic surfaces in the system only up to tens of nanometers.*

Fluids, the flow of liquid or gas, have properties that vary continuously under the action of external forces. In the presence of fluid, shear forces are small in magnitude, which leads to large changes in the relative position of the element of fluid. In contrast, changes in the relative positions of atoms in solids remain small under the action of any small external force. Termination of action of the external forces on the fluid does not necessarily lead to the restoration of its initial form.

7.2.1   CAPILLARY EFFECTS

To observe the capillary effects, one must open the nanotube, that is, to re-move the upper part lids. Fortunately, this operation is quite simple.

The first study of capillary phenomena has shown that there is a relation-ship between the magnitude of surface tension and the possibility of its be-ing drawn into the channel of the nanotube. It was found that the liquid penetrates into the channel of the nanotube, if its surface tension is not higher than 200 mN/m. For example, concentrated nitric acid with surface tension of 43 mN/m is used to inject certain metals into the channel of a nanotube. Then annealing is conducted at 4,000°C for 4 hours in an atmosphere of hydrogen, which leads to the recovery of the metal.

Along with the metals, CNTs can be filled with gaseous substances, such as hydrogen in molecular form. This ability is of great practical importance and can be used as a clean fuel in internal combustion engines.

7.2.2 SPECIFIC ELECTRICAL RESISTANCE OF CNTs (ρ)

The resistivity of the nanotubes can be varied within wide limits to 0.8 ohm/ cm. The minimum value is lower than that of graphite. Most of the nanotubes have metallic conductivity, and the smaller nanotubes show properties of a semiconductor with a band gap of 0.1–0.3 eV.

The resistance of single-walled nanotube is independent of its length, because of this it is convenient to use for the connection of logic elements in microelectronic devices. The permissible current density in CNTs is much greater than in metallic wires of the same cross-section and one hundred times better achievement for superconductors.

7.2.3 EMISSION PROPERTIES OF CNTs

The results of the study of emission properties of the material (where the nanotubes were oriented perpendicular to the substrate) have been very interesting for practical use. An attained value of the emission current density is of the order of 0.5 mA/mm^2. The value obtained is in good agreement with the Fowler–Nordheim expression.

7.2.4 ELECTROKINETIC PROCESSES IN MICRO- AND NANOSCALE SYSTEMS

The most effective and common way to control microflow substances are *electrokinetic* and *hydraulic*. At the same time, the most technologically advanced and automated considered electrokinetic.

Charge transfer in mixtures occurs as a result of the directed motion of charge carrier ions. There are different mechanisms of such transfer, but usually are *convection, migration,* and *diffusion.*

Convection is called mass transfer of the macroscopic flow. *Migration* is the movement of charged particles by electrostatic fields. The velocity of the ions depends on field strength. In microfluidics, the *electrokinetic process* can be divided into *electro-osmosis, electrophoresis, streaming potential,* and *sedimentation potential.* These processes can be qualitatively described as follows:

(a) *Electro-osmosis*: This is movement of the fluid volume in response to the applied electric field in the channel of the electrical double layers on its wetted surfaces.

(b) *Electrophoresis*: This is forced motion of charged particles or molecules, in mixture with the acting electric field.

(c) *Streamy potential*: This is electric potential, which is distributed through a channel with charged walls, in the case when the fluid moves under the action of pressure forces. Joule electric current associated with the effect of charge transfer is flowing stream.

(d) *The potential of sedimentation*: An electric potential is created when charged particles are in motion relative to a constant fluid.

In general, for the microchannel cross-section, S, amount of introduced probe (when entering electrokinetic method) depends on the applied voltage, U, time, t, during which the received power, and mobility of the sample components, μ:

$$Q = \frac{\mu S U t}{L} \cdot c$$

where

c, probe concentration in the mixture

L, the channel length.

Amount of injected substance is determined by the electrophoretic and total electroosmotic mobilities μ.

In the hydrodynamic mode of entry by the pressure difference in the channel or capillary of circular cross-section, the volume of injected probe, V_c:

$$V_c = \frac{4}{128} \cdot \frac{\Delta p \pi d t}{\eta L}$$

where
    Δp, pressure differential
    d, diameter of the channel
    η, viscosity

7.2.5 CONTINUUM HYPOTHESIS

In the simulation of processes in micron-sized systems ,the following basic principles are fundamentals:

(a) Hypothesis of *laminar* flow (sometimes is taken for granted when it comes to microfluidics)
(b) Continuum hypothesis (detection limits of applicability)
(c) Laws of formation of the velocity profile, mass transfer, and distribution of electric and thermal fields
(d) Boundary conditions associated with the geometry of structural elements (walls of channels, mixers zone flows, etc.)

Because we consider the physical and chemical transport processes of matter and energy, mathematical models have the form of systems of differential equations of second-order partial derivatives. Methods for solving such equations are analytical (Fourier and its modifications, such as the method of Greenberg, Galerkin, the method of d'Alembert and the Green's functions, the Laplace operator method, etc.) or numerical (explicit or more effectively, implicit finite difference schemes).

Laminar flow: This is a condition in which the particle velocity in the liquid flow is not a random function of time. The small size of the microchannels (typical dimensions of 5–300 μm) and low surface roughness create good conditions for the establishment of laminar flow. Traditionally, the image of the nature of the flow gives the dimensionless characteristic numbers: Reynolds number and Darcy's friction factor.

In the motion of fluids in channels, the turbulent regime is rarely achieved. At the same time, the movement of gases is usually turbulent.

Although the liquids are quantized in the length scale of intermolecular distances (about 0.3–3 nm in liquids and for gases), they are assumed to be continuous in most cases. Continuum hypothesis (continuity, continuum) suggests that the macroscopic properties of fluids consist of molecules, the same as if the fluid were completely continuous (structurally homogeneous). Physical characteristics such as mass, momentum, and energy associated with the volume of fluid containing a sufficiently large number of molecules must be taken as the sum of all the relevant characteristics of the molecules.

Continuum hypothesis leads to the concept of fluid particles. In contrast to the ideal of a point particle in ordinary mechanics, in fluid mechanics, particle in the fluid has a finite size.

At the atomic scale, there are large fluctuations due to the molecular structure of fluids; but if we increase the sample size, we reach a level where it is possible to obtain stable measurements. This volume of probe must contain a sufficiently large number of molecules to obtain reliable reproducible signal with small statistical fluctuations. For example, if we determine the required volume as a cube with sides of 10 nm, this volume contains some of the molecules and determines the level of fluctuations of the order of 0.5%.

The applicability of the hypothesis is based on comparison of free path length of a particle λ in a liquid with a characteristic geometric size d. The ratio of these lengths is called the Knudsen number: $Kn = \lambda / d$.

When (1) $Kn < 10^{-3}$ justifies hypothesis of a continuous medium and when (2) $Kn < 10^{-1}$ allows the use of adhesion of particles to the solid walls of the channel.

Theoretical condition can also be varied: both in form $U = 0$ and in a more complex form, associated with shear stresses. The calculation of λ can be carried out as $\lambda \approx \sqrt[3]{\overline{V} / Na}$,

where

$\overline{V}$, molar volume

Na, Avogadro's number

Under certain geometrical approximations of the particles of substance, free path length can be calculated as $\lambda \approx 1 / \left(\sqrt{2} \pi r_S^2 Na \right)$ (if used instead r_S Stokes radius, as a consequence of the spherical approximation of the particle). On the contrary, for a rigid model of the molecule r_S should be replaced by the characteristic size of the particles R_g (the radius of inertia), calculated as $R_g = n_i \cdot \delta_1 / \sqrt{6}$.

where

δ_1, the length of a fragment of the chain (link)

n_i, the number of links.

Of course, the continuum hypothesis is not acceptable when the system under consideration is close to the molecular scale. This happens in nanoliquid, such as liquid transport through *nano*-pores in cell membranes or artificially made nanochannels.

7.2.6   THE MOLECULAR DYNAMICS METHOD

In contrast to the continuum hypothesis, the essence of modeling the molecular dynamics method is as follows. We consider a large ensemble of particles that simulate atoms or molecules, that is, all atoms are material points. It is believed that the particles interact with each other and, moreover, may be subject to external influence. Interatomic forces are represented in the form of the classical potential force (the gradient of the potential energy of the system).

The interaction between atoms is described by means of van der Waals forces (intermolecular forces), mathematically expressed by the Lennard–Jones potential:

$$V(r) = \frac{Ae^{-\sigma r}}{r} - \frac{C_6}{r^6}$$

where

A and C_6 are some coefficients depending on the structure of the atom or molecule and σ is the smallest possible distance between the molecules.

In the case of two isolated molecules at a distance of r_0, the interaction force is zero (i.e., the repulsive forces balance attractive forces). When $r > r_0$ the resultant force is the force of gravity, which increases in magnitude, reaching a maximum at $r = r_m$ and then decreases. When $r < r_0$ there is a repulsive force. Molecule in the field of these forces has potential energy $V(r)$, which is connected with the force of $f(r)$ by the differential equation

$dV = -f(r)dr$

At the point where $r = r_0$, $f(r) = 0$, $V(r)$ reaches an extremum (minimum).

The chart of such a potential is as shown in Figure 24. The upper (positive) half-axis r corresponds to the repulsion of the molecules and the lower (negative) half-plane shows their attraction. One can observe that at short distances the molecules predominantly repel each other.

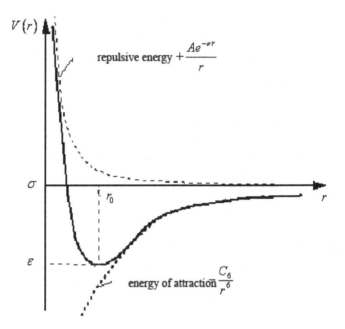

FIGURE 24   Potential energy of intermolecular interaction.

The exponential summand in the expression for the potential describing the repulsion of the molecules at small distances is often approximated as follows: $\dfrac{Ae^{-\sigma r}}{r} \approx \dfrac{C_{12}}{r^{12}}$

In this case, we obtain the Lennard–Jones potential:

$$V(r) = \frac{C_{12}}{r^{12}} - \frac{C_6}{r^6} \qquad (2.1)$$

The interaction between carbon atoms is described by the potential

$$V_{CC}(r) = K(r - b)^2,$$

where

K = constant tension (compression) connection

$b = 1.4A$, the equilibrium length of connection

r, current length of the connection

The interaction between the carbon atom and hydrogen molecule is described by the Lennard–Jones potential

$$V(r) = 4\varepsilon\left[\left(\frac{\sigma}{r}\right)^{12} - \left(\frac{\sigma}{r}\right)^6\right]$$

For all particles (Figure 25), the equations of motion are given by the following equation:

$$m\frac{d^2\overline{r}_i}{dt^2} = \overline{F}_{T-H_2}(\overline{r}_i) + \sum_{j \neq i}\overline{F}_{H_2-H_2}(\overline{r}_i - \overline{r}_j)$$

,

where
$\overline{F}_{T-H_2}(\overline{r})$, force acting by the CNT

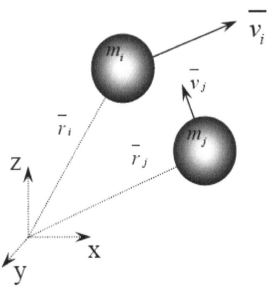

FIGURE 25 $\overline{F}_{H_2-H_2}(\overline{r}_i - \overline{r}_j)$; Force acting on the i-th molecule from the j-th molecule

The resulting system of equations is solved numerically. However, the molecular dynamics method has limitations of applicability:

(a) the de Broglie wavelength h/mv

where

h, Planck's constant

m, the mass of the particle

v, velocity

(b) Classical molecular dynamics cannot be applied for modeling systems consisting of light atoms such as helium or hydrogen

(c) At low temperatures, quantum effects become decisive for the consideration of such systems and must use quantum chemical methods

(d) The time at which we consider the behavior of the system is more than the relaxation time of the physical quantities

The coordinates of the molecules are distributed regularly in space. The velocities of the molecules are distributed according to the Maxwell equilibrium distribution function according to the temperature of the system:

$$f(u,v,w) = \frac{\beta^3}{\pi^{3/2}} \exp\left(-\beta^2\left(u^2+v^2+w^2\right)\right) \quad \beta = \frac{1}{\sqrt{2RT}}$$

The macroscopic flow parameters are calculated from the distribution of positions and velocities of the molecules:

$$\overline{V} = \left\langle \overline{v}_i \right\rangle = \frac{1}{n}\sum_i \overline{v}_i, \quad \frac{3}{2}RT = \frac{1}{2}\left\langle \left|\overline{v}_i'\right|^2 \right\rangle, \quad \overline{v}_i' = \overline{v}_i - \overline{V},$$

$$\rho = \frac{nm}{V_0},$$

7.2.7 VAN DER WAALS, EQUATION. CORRESPONDING STATES LAW

In 1873, Van der Waals proposed an equation of state is qualitatively good description of liquid and gaseous systems. For one mole, the equation is

$$\left(p + \frac{a}{v^2}\right)(v - b) = RT \tag{2.2}$$

Note that at $p > \dfrac{a}{v^2}$ and $v > b$, this equation becomes the equation of state of ideal gas

$$pv = RT \qquad (2.3)$$

Van der Waals equation can be obtained from the Clapeyron equation of Mendeleev by an amendment to the magnitude of the pressure a/v^2 and the amendment b to the volume, both constant a and b independent of T and v but dependent on the nature of the gas.

The amendment b takes the following into account:

(a) the volume occupied by the molecules of real gas (in an ideal gas molecules are taken as material points, not occupying any volume);

(b) so-called dead space, that cannot penetrate the molecules of real gas during motion, that is, volume of gaps between the molecules in their dense packing.

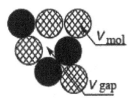

FIGURE 26 Location scheme of molecules in a real gas.

Thus, $b = v_{MO\pi} + v_{3a3}$ (Figure 26). The amendment a/v^2 takes into account the interaction force between the molecules of real gases. It is the internal pressure, which is determined from the following simple considerations. Two adjacent elements of the gas will react with a force proportional to the product of the quantities of substances enclosed in these elementary volumes.

TABLE 3 The molecular weight of some gases

Gas	N	Ar	H_2	O_2	O	CO_2	Ammonia	Air
μ	28	40	2	32	28	44	17	29

Therefore, the internal pressure p_{BH} is proportional to the square of the concentration n:

$$p_{\text{вн}} \sim n^2 \sim \rho^2 \sim \frac{1}{v^2},$$

where ρ is the gas density.

Thus, the total pressure consists of internal and external pressures:

$$p + p_{\text{вн}} = p + \frac{a}{v^2}$$

Equation (2.3) is the most common for an ideal gas.

Under normal physical conditions ($p_0 = 0.1013 М\Pi a$, $t_0 = 0^o C$) $v\mu = 22,4 \, м^3 / (кмоль \times K^o)$, and then from Eq. (2.3):

$$R\mu = \frac{pv\mu}{T} = \frac{0,1013 \cdot 10^6 \cdot 22,4}{273} = 8314 \, \frac{дж}{кмоль \cdot K^0}$$

Knowing $R\mu$ we can find the gas constant for any gas with the help of the value of its molecular mass μ (Table 3):

$$R = \frac{R\mu}{\mu} = \frac{8314}{\mu}$$

For gas mixture with mass M state equation has the form

$$pv = MR_{CM}T = \frac{8314MT}{\mu_{CM}} \qquad (2.4)$$

where R_{CM} is gas constant of the mixture.

The gas mixture can be given by the mass proportions g_i, voluminous r_i, or mole fractions n_i respectively, which are defined as the ratio of mass m_i, volume v_i or number of moles N_i of i gas to total mass M, volume v or number of moles N of gas mixture. Mass fraction of component is $g_i = \frac{m_i}{M}$, where $i = 1, n$. It is obvious that $M = \sum_{i=1}^{n} m_i$ and $\sum_{i=1}^{n} g_i = 1$. The volume fraction is $r_i = \frac{v_i}{v_{CM}}$, where v_i is the partial volume of component mixtures.

Similarly, $\sum_{i=1}^{n} v_i = v_{CM}, \sum_{i=1}^{n} r_i = 1$.

Depending on specificity of tasks, the gas constant of the mixture can be determined as follows:

$$R_{CM} = \sum_{i=1}^{n} g_i R_i \; ; \quad R_{CM} = \frac{1}{\sum_{i=1}^{n} r_i R_i^{-1}}$$

If we know the gas constant R_{CM}, the seeming molecular weight of the mixture is equal to

$$\mu_{CM} = \frac{8314}{R_{CM}} = \frac{8314}{\sum_{i=1}^{n} g_i R_i} = 8314 \sum_{i=1}^{n} r_i R_i^{-1}$$

The pressure of the gas mixture p is equal to the sum of the partial pressures of individual components in the mixture p_i:

$$p = \sum_{i=1}^{n} p_i \tag{2.5}$$

Partial pressure p_i is pressure of gas; if it is one, at the same temperature fills the whole volume of the mixture ($p_i v_{CM} = RT$).

With various methods of setting the gas mixture partial pressures

$$p_i = p r_i \; ; \quad P_i = \frac{p g_i \mu_{CM}}{\mu_i} \tag{2.6}$$

From the expression Eq (2.6), we determine that for the calculation of the partial pressures p_i it is necessary to obtain the pressure of the gas mixture, the volume or mass fraction i of the gas component, as well as the molecular weight of the gas mixture μ and the molecular weight of i of gas μ_i.

The relationship between mass and volume fractions are as follows:

$$g_i = \frac{m_i}{m_{CM}} = \frac{\rho_i v_i}{\rho_{CM} v_{CM}} = \frac{R_{CM}}{R_i} r_i = \frac{\mu_i}{\mu_{CM}} r_i$$

Equation (2.2) can be rewritten as follows:

$$v^3 - \left(b + \frac{RT}{p}\right)v^2 + \frac{a}{p}v - \frac{ab}{p} = 0 \tag{2.7}$$

When $p = p_k$ and $T = T_k$, where p_k and T_k is the critical pressure and temperature, all three roots of Eq (2.7) are equal to the critical volume v_k

$$v^3 - \left(b + \frac{RT_k}{P_k}\right)v^2 + \frac{a}{p_k}v - \frac{ab}{p_k} = 0 \tag{2.8}$$

Because $v_1 = v_2 = v_3 = v_k$, then Eq (2.8) must be identical to the equation

$$(v - v_1)(v - v_2)(v - v_3) = (v - v_k)^3 = v^3 - 3v^2 v_k + 3vv_k^2 - v_k^3 = 0 \tag{2.9}$$

Comparing the coefficients at the equal powers of v in both equations leads to the equalities

$$b + \frac{RT_k}{p_k} = 3v_k \, ; \, \frac{a}{p_k} = 3v_k^2 \, ; \, \frac{ab}{p_k} = v_k^3 \tag{2.10}$$

Hence,

$$a = 3v_k^2 p_k \, ; \, b = \tfrac{v_k}{3} \tag{2.11}$$

Considering Eq (2.10) as equation for the unknowns p_k, v_k, T_k,

$$p_k = \frac{a}{27b^2} \, ; \, v_k = 3b \, ; \, T_k = \frac{8a}{27bR} \tag{2.12}$$

From Eqs (2.10) and (2.11) or (2.12), we can find the relation

$$\frac{RT_k}{p_k v_k} = \frac{8}{3} \tag{2.13}$$

Instead of the variables p, v, T let us introduce the relationship of these variables to their critical values (leaden dimensionless parameters)

$$\pi = \frac{p}{p_k}; \quad \omega = \frac{v}{V_k}; \quad \tau = \frac{T}{T_k} \tag{2.14}$$

Substituting Eqs (2.12) and (2.14) in (2.7) and using Eq (2.13)

$$\left(\pi p_k + \frac{3v_k^2 p_k}{\omega^2 v_k^2}\right)\left(\omega v_k - \frac{v_k}{3}\right) = RT_k\tau$$

$$\left(\pi + \frac{3}{\omega^2}\right)(3\omega - 1) = 3\frac{RT_k}{p_k v_k}\tau$$

$$\left(\pi + \frac{3}{\omega^2}\right)(3\omega - 1) = 8\tau \tag{2.15}$$

In Eq. (2.15), a and b are not permanent, depending on the nature of the gas. That is, if the units of measurement of pressure, volume, and temperature are used as their critical values (use the leaden parameters), the equation of state is the same for all substances.

This condition is called the *law of corresponding states.*

7.3 SLIPPAGE OF THE FLUID PARTICLES NEAR THE WALL

According to the Navier boundary condition, the velocity slip is proportional to fluid velocity gradient at the wall:

$$v\big|_{y=0} = L_S \, dv/dy\big|_{y=0} \tag{3.1}$$

Here and in Figure 27,

L_S represents the "slip length" and has a dimension of length.

Because of the slippage, the average velocity in the channel $\langle v_{pdf} \rangle$ increases.

In a rectangular channel (of width >> height h and viscosity of the fluid η) due to an applied pressure gradient of dp/dx:

$$\langle v_{pdf} \rangle = \frac{h^2}{12\eta}\left(-\frac{dp}{dx}\right)\left(1+\frac{6L_S}{h}\right)$$

(3.2)

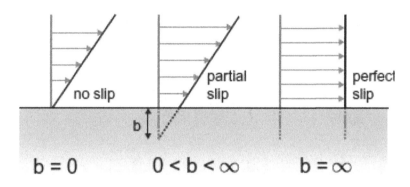

FIGURE 27   Three cases of slip flow (with slip length b).

The results of molecular dynamics simulation for nanosystems with liquid show that large slippage lengths (of the order of microns) should occur in the CNTs of nanometer diameter and, consequently, can increase the flow rate by three orders of ($6L_S/h > 1000$). Thus, the flow with slippage is becoming more and more important for hydrodynamic systems of small size.

The results of molecular dynamics simulation of unsteady flow of mixtures of water—water vapor, water, and nitrogen in a CNT—were reported previously. Based on these studies, a flow of water through CNTs with different diameters at temperature of $300°K$ was considered.

CNTs have been considered "zigzag" with chiral vectors (20, 20), (30, 30), and (40, 40), corresponding to pipe diameters or 2.712, 4.068, and 5.424 nm, respectively.

The value of the flow rate and the system pressure, which varies in the range of 600–800 bars, were high enough to ensure complete filling of the tubes. This pressure was achieved by the total number of water molecules 736, 904, and 1694.

The effects of slippage of various liquids on the surface of the nanotube were studied in detail.

The lengths of slip, calculated using the current flow velocity profiles of liquid, are shown in Figure 28, are 11, 13, and 15 nm for the pipes of 2.712, 4.068, and 5.424 nm, respectively. The dotted line is marked by theoretical modeling data. The vertical lines indicate the position of the surface of CNTs.

It was found that as the diameter decreases, the speed of slippage of particles on the wall of nanotube also decreases. The report attributes this to the increase of the surface friction.

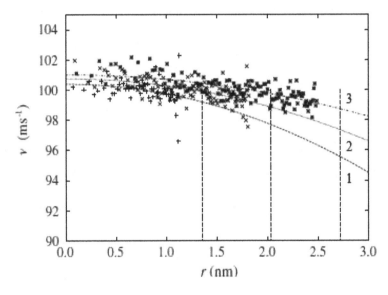

FIGURE 28   Time average streaming velocity profiles of water in a CNTs of different diameter: 2.712 nm, curve 1; 4.068 nm, curve 2; 5.424 nm, curve 3.

Experiments with various pressure drops in nanotubes demonstrated slippage of fluid in micro- and nanosystems. The most remarkable were the two recent experiments, which were conducted to improve the flow characteristics of CNTs with the diameters of 2 and 7 nm.

In the membranes in which the CNTs were arranged in parallel, there was a slip of the liquid in the micrometer range. This led to a significant increase in flow rate, up to three and four orders of magnitude.

Experiments for the water moving in microchannels on smooth hydrophobic surfaces show a sliding at about 20 nm. If the wall of the channel is not smooth but twisty or rough (and hydrophobic) such a structure would lead to an accumulation of air in the cavities and become superhydrophobic (with contact angle greater than $160°$). It is believed that this leads to the creation of contiguous areas with high and low slippage, which can be described as "effective slip length." This effective length of the slip occurring on the rough surface can be several tens of microns, which was confirmed experimentally.

It should be noted that for practical use, the advantages of nanotubes with slippage is necessary for solving various problems. Scientists have already shown that hydrophobic surfaces tend to form bubbles. On the contrary, the surfaces used by most researchers were rough, but the use of smooth surfaces could generally reduce the formation of bubbles.

Another possible problem is filling of the hydrophobic systems with liquid. Filling of micron-sized hydrophobic capillaries is not a big problem, because pressure of less than 1 atm is sufficient. Capillary pressure, however, is inversely proportional to the diameter of the channel, and filling for nanochannels can be very difficult.

7.3.1   THE DENSITY OF THE LIQUID LAYER NEAR A WALL OF CNT

Scientists showed radial density profiles of oxygen averaged in time and hydrogen atoms in the "zigzag" CNT with chiral vector (20,20) and a radius $R = 1.356$ nm (Figure 29). The distribution of molecules in the area near the wall of the CNT indicated a high-density layer near the wall of the CNT. Such a pattern indicates the presence of structural heterogeneity of the liquid in the flow of the nanotube. In Figure 29, $\rho^*(r) = \rho(r)/\rho(0)$, whereas 2.712 nm diameter pipe is completely filled with water molecules at $300°K$. The overall density is $\rho^*(r) = \rho(r)/\rho(0)$. The arrows denote the location of distinguishable layers of the water molecules and the vertical line is the position of the CNT wall.

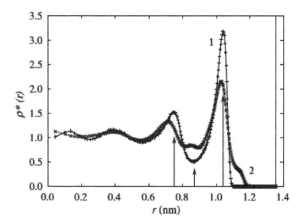

FIGURE 29 Radial density profiles of oxygen (curve 1) and hydrogen (curve 2) atoms.

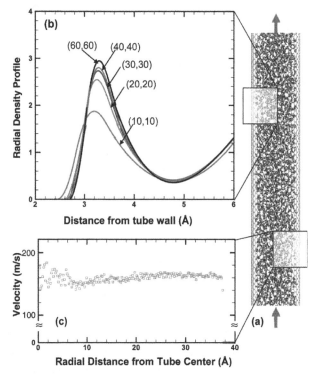

FIGURE 30 Initial structure and movement of water molecules: (a) schematic of the initial structure and transport of water molecules in a model CNT, (b) the radial density profile (RDP) of water molecules inside CNTs with different radii, and (c) the representative radial velocity profile (RVP) of water molecules inside a (60,60) nanotube.

The distribution of molecules in the region of $0.95 \leq r \leq R$ nm indicates a high-density layer near the wall CNT. Such a pattern indicates the presence of structural heterogeneity of the liquid in the flow of the nanotube.

Figure 30 shows the scheme of the initial structure and movement of water molecules in

(a) the model CNT

(b) the radial density profile of water molecules inside nanotubes with different radii

(c) the velocity profile of water molecules in a nanotube chirality (60,60)

It should be noted that the available area for molecules of the liquid is less than the area bounded by a solid wall, primarily due to the Van der Waals interactions.

7.3.2 THE EFFECTIVE VISCOSITY OF THE LIQUID IN A NANOTUBE

There is a significant increase in the effective viscosity of the fluid in the nanovolumes compared with its macroscopic value. However, the effective viscosity of the liquid in a nanotube depends on the diameter of the nanotube.

The effective viscosity of the liquid in a nanotube is defined as follows.

Conformity nanotubes filled with liquid, containing crystallites with the same size tube filled with liquid, can be considered as a homogeneous medium (i.e., without considering the crystallite structure), in which the same pressure drop and flow rate of Poiseuille flow is realized. The viscosity of a homogeneous fluid, which ensures the coincidence of these parameters, is called the effective viscosity of the flow in the nanotube.

While flowing in the narrow channels of width less than 2 nm, water behaves like a viscous liquid. In the vertical direction, water behaves as a rigid body, and in a horizontal direction it maintains its fluidity.

It is known that at large distances the van der Waals, interaction has a magnetic tendency and occurs between any molecules such as polar as well as nonpolar. At small distances, it is compensated by repulsion of electron shells. Van der Waals, interaction decreases rapidly with distance. Mutual convergence of the particles under the influence of magnetic forces continues until these forces are balanced with the increasing forces of repulsion.

Knowing the deceleration of the flow (Figure 30) of water a, the effective shear stress between the wall of the pipe length l and water molecules can be calculated by

$$\tau = Nma/(2\pi Rl) \tag{3.3}$$

where the shear stress is a function of tube radius and flow velocity $\bar{v}$ and m is mass of water molecules. The average speed is related to volumetric flow $\bar{v} = Q/(\pi R^2)$.

Denoting n_0 the number of water molecules, we can calculate the shear stress in the form

$$\tau\big|_{r=R} = n_0 mRa/2 \tag{3.4}$$

Figure 31 shows the results of calculations concerning the influence of the size of the tube R_0 on the effective viscosity (squares) and shear stress τ (triangles), when the flow rate is approximately 165 m/sec.

According to classical mechanics of liquid flow at different pressure drops Δp along the tube length l is given by Poiseuille formula:

$$Q_P = \frac{\pi R^4 \Delta p}{8 \eta l}, \tag{3.5}$$

Therefore,

$$\tau = \frac{\Delta p R}{2l} \tag{3.6}$$

and the effective viscosity of the fluid can be estimated as $\eta = \tau \cdot R/(4\bar{v})$.

The change in the value of shear stress directly causes the dependence of the effective viscosity of the fluid from the pipe size and flow rate. In this case, the effective viscosity of the transported fluid can be determined from Eqs (3.5) and (3.6) as follows

$$\eta = \frac{\tau \cdot R}{4\bar{v}} \tag{3.7}$$

It should be noted that the magnitude of the shear stress τ is relatively small in the range of pipe sizes considered. This indicates that the surface of CNTs is very smooth and the water molecules can easily slide through it.

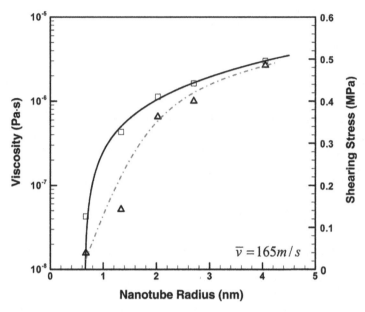

FIGURE 31   Size effect of shearing stress (triangle) and viscosity (square).

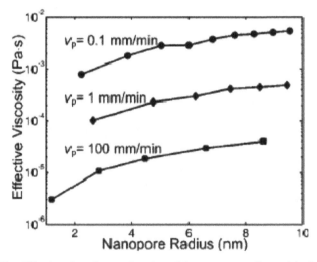

FIGURE 32   Effective viscosity as a function of the nanopore radius and the loading rate.

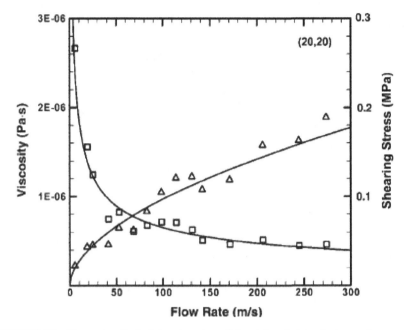

FIGURE 33   Flow rate effect of shearing stress (triangle) and viscosity (square).

In fact, shear stress is primarily due to van der Waals, interaction between the solid wall and the water molecules. It is noted that the characteristic distance between the near-wall layer of fluid and pipe wall depends on the equilibrium distance between atoms O and C and the distribution of the atoms of the solid wall and bend of the pipe.

From Figure 31, the effective viscosity η increases by two orders of magnitude when R_0 ranging from 0.67 to 5.4 nm.

According to Eqs (3.5)–(3.7), the effective viscosity can be calculated as

$\eta = \frac{\pi R^4 \Delta p}{8QL}$. The results of calculations are shown in Figure 32.

The dependence of the shear stress on the flow rate is illustrated in Figure 33. For the tube (20,20) τ increases with v. The growth rate slowed down at higher values v.

At high speeds v, while water molecules are moving along the surface of the pipe, the liquid molecules do not have enough time to fully adjust their positions to minimize the free energy of the system. Therefore, the distance between adjacent carbon atoms and water molecules may be less than the equilibrium van der Waals, distances. This leads to an increase in van der Waals forces of repulsion and leads to higher shear stress.

It should be noted that even though the equation for viscosity is based on the theory of the continuum, it can be extended to a complex flow to determine the effective viscosity of the nanotube.

Figure 33 shows dependence of η on $\bar{v}$ inside of the nanotube (20, 20). It is shown that η decreases sharply with increasing flow rate and begins to approach a definite value when $v > 150$ m/sec. For the current pipe size and flow rate ranges of $\eta \sim 1/\sqrt{\bar{v}}$, this trend is because of $\tau - v$, contained in Figure 33 According to Figure 32 high-speed effects are negligible.

One can easily observe that the dependence of viscosity on the size and speed is consistent qualitatively with the results of molecular dynamic simulations. In all the studied cases, the viscosity is much smaller than its macroscopic analogy. As the radius of the pores varies from about 1 to 10 nm, the value of the effective viscosity increases by an order of magnitude. A more significant change occurs when increasing the speed of 0.1 mm/min up to 100 mm/min. This results in a change in the value of viscosity η, respectively, by 3–4 orders. The discrepancy between simulation and test data can be associated with differences in the structure of the nanopores and liquid phase.

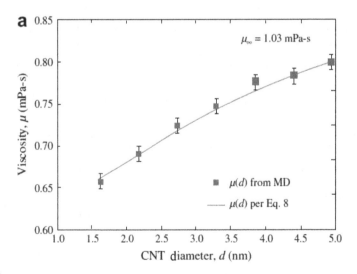

FIGURE 34   Variation of water viscosity with CNT diameter.

Figure 34 shows the viscosity dependence of water (calculated by the method of DM and the diameter of the CNT). The viscosity of water, as shown in the figure, increases monotonically with increasing diameter of the CNT.

7.3.3 THE RELEASE OF ENERGY DUE TO THE COLLAPSE OF THE NANOTUBE

Scientists theoretically predicted the existence of a "domino effect" in single-walled CNT.

Squashing can occur at one end by two rigid movements of narrow graphene planes (about 0.8 nm in width and 8.5 nm in length). This can rapidly (at a rate exceeding 1 km/s) release its stored energy by collapsing along its length like a row of dominoes. The effect resembles a tube of toothpaste squeezing itself (Figure 35).

The structure of a single-walled CNT has two possible stable states: circular or collapsed. Scientists realized that for nanotubes wider than 3.5 nm, the circular state stores more potential energy than the collapsed state as a result of van der Waal's forces. He performed molecular dynamics simulations to find out what would happen if one end of a nanotube was rapidly collapsed by clamping it between two graphene bars.

This phenomenon occurs with the release of energy, and thus allows for the first time to consider CNTs as energy sources. This effect can also be used as an accelerator of molecules.

The tube collapses at the same time not over its entire length, and sequentially, one after the other carbon ring, starting from the end, which is tightened (Figure 35). It happens just like a domino collapses, arranged in a row (this is known as the "domino effect"). The role of bone dominoes performing here as a ring of carbon atoms form the nanotube, and the nature of this phenomenon is quite different.

Recent studies show that for nanotubes with diameters ranging from 2 to 6 nm, there are two stable equilibrium states:
-   cylindrical (tube no collapses) and
-   compressed (imploded tube) with different values of potential energy (the difference between which can be used as an energy source).

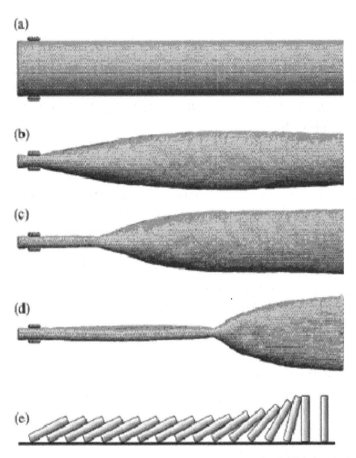

FIGURE 35   "Domino effect" in a carbon nanotube: (a) the initial form of carbon nanotubes is cylindrical; (b) one end of the tube is squeezed; (c), (d) propagation of domino waves; the configuration of the nanotube 15 and 25 picoseconds after the compression of its end; and (e) schematic illustration of the "domino effect" under the influence of gravity.

The switching between these two states with the subsequent release of energy occurs in the form of waves arising domino effect. The scientists have shown that such switching is carried out in CNTs with diameters of 2 nm.

A theoretical study of the "domino effect" was conducted by using a special method of classical molecular dynamics, in which the interaction between carbon atoms was described by van der Waals, forces.

The main reason for the observed effect is the potential energy of the van der Waals interactions (which "collapses" the nanotube) and the energy of elastic deformation (which seeks to preserve the geometry of carbon atoms). This eventually leads to a bistabled (collapsed and no collapsed) configuration of CNT.

Thus, "domino effect" wave can be produced in a CNT with a relatively large diameter (more than 3.5 nm), because only in such a system the potential energy of collapsing structures may be less than the potential energy of the "normal" nanotube. In other words, the cylindrical structure and collapsing nanotubes with large diameters are, respectively, of the metastable and stable states.

The potential energy of a CNT with a propagating wave at the "domino effect" is as shown in Figure 35(a).

This figure shows three sections:

The first (from 0 ps to 10 ps) are composed of elastic strain energy, which appears due to changes in the curvature of the walls of the nanotubes in the process of collapsing.

The second region (from 10 ps to 35 ps) corresponds to the "domino effect."

Finally, the third segment (from 35 ps to 45 ps) corresponds to the process is ended "domino."

The spread "domino effect" waves are a process that goes with the release of energy (about 0.01 eV per atom of carbon). This is certainly not comparable in anyway with the degree of energy yield in nuclear reactions, but the fact of power generation CNT is obvious.

Calculations show that the wave of dominoes in a tube with diameter of 4–5 nm is about 1 km/s (as seen from the Figure 35(b)) and in a nonlinear manner depending on its geometry (the diameter and chirality). The maximum effect should be observed in the tube with a diameter slightly less than 4.5 nm (considering carbon rings to collapse at a speed of 1.28 km/sec). The theoretical dependence shows the blue solid line. And now an example of how energy is released in such a system with a "domino effect" can be used in nanodevices. Scientists offer an original way to use a sort of "nanogun" (Figure 36(a)). Imagine that at our disposal is a CNT with chirality (55.0) and similar to the dominoes in diameter. Put inside a nanotube fullerene C_{60}. With a little imagination, this can be considered a CNT as the gun trunk, and the molecule (as its shell).

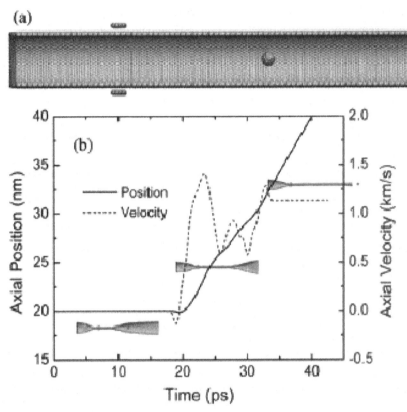

FIGURE 36   Nanocannon scheme acting on the basis of "domino effect" in the incision: (a) inside a CNT (55.0) is the fullerene molecule C_{60} and (b) the initial position and velocity of the departure of the "core" (a fullerene molecule), depending on the time. The highest rate of emission of C_{60} (1.13 km/sec) comparable to the velocity of the domino wave.

The question is what is the speed of the "core?" Based on the initial position of the fullerene molecule on departure of a nanotube, it can reach speeds close to the velocity of "domino effect" waves, which is about 1 km/sec (Figure 36(b)). Interestingly, this speed is reached by the "core" for just 2 ps and at a distance of 1 nm. It is easy to calculate that the observed acceleration is of great value $0.5 \cdot 10^{15} \, M/c^2$.

7.4 FLUID FLOW IN NANOTUBES

Scientists from the University of Wisconsin–Madison (USA) managed to prove that the laws of friction for the nanostructures do not differ from the classical laws.

The friction of surface against the surface in the absence of the interlayer between the liquid materials (so-called dry friction) is created by irregularities in the given surfaces that rub one another, as well as the interaction forces between the particles that make up the surface.

As part of their study, the researchers built a computer model that calculates the friction force between nanosurfaces (Figure 37). In the model, these surfaces were presented simply as a set of molecules for which forces of intermolecular interactions were calculated.

As a result, scientists were able to establish that the friction force is directly proportional to the number of interacting particles. The researchers propose to consider this quantity by analog of so-called true macroscopic contact area. It is known that the friction force is directly proportional to this area (it should not be confused with common area of the contact surfaces of the bodies).

In addition, the researchers were able to show that the friction surface of the nanosurfaces can be considered within the framework of the classical theories of friction of nonsmooth surfaces.

A literature review shows that nowadays molecular dynamics and mechanics of the continuum are the main methods of research of fluid flow in nanotubes.

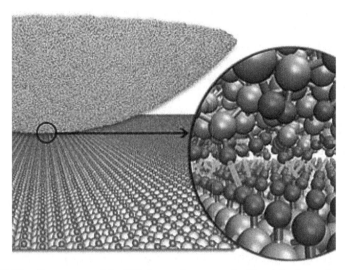

FIGURE 37   Computer model of friction at the nanoscale. The right shows the surfaces of interacting particles

Although the method of molecular dynamics simulations is effective, at the same time it requires enormous computing time especially for large systems. Therefore, simulation of large systems is more reasonable to carry out nowadays by the method of continuum mechanics [5–11].

In the work of Morten Bo Lindholm Mikkelsen et al. [12], the fluid flow in the channel is considered in the framework of the continuum hypothesis. The Navier–Stokes equation was used and the velocity profile was determined for Poiseuille flow.

In the work of Thomas John A. and McGaughey Alan JH [13], the water flow by means of pressure differential through the CNTs with diameters ranging from 1.66 to 4.99 nm is researched using molecular dynamics simulation study. For each nanotube, the value enhancement predicted by the theory of liquid flow in the CNTs is calculated. This formula is defined as a ratio of the observed flow in the experiments to the theoretical values without considering slippage on the model of Hagen–Poiseuille. The calculations showed that the enhancement decreases with increasing diameter of the nanotube.

Important conclusion of the Thomas John A. and McGaughey Alan JH [13] is that by constructing a functional dependence of the viscosity of the water and length of the slippage on the diameter of CNTs, the experimental results in the context of continuum fluid mechanics can easily be described. The aforementioned is true even for CNTs with diameters of less than 1.66 nm.

The theoretical calculations use the following formula for the steady velocity profile of the viscosity η of the fluid particles in the CNT under pressure gradient $\partial p / \partial z$:

$$v(r) = \frac{R^2}{4\eta}\left[1 - \frac{r^2}{R^2} + \frac{2L_s}{R}\right]\frac{\partial p}{\partial z} \tag{4.1}$$

The length of the slip, which expresses the speed heterogeneity at the boundary of the solid wall and fluid, is defined as[14–16]

$$L_S = \frac{v(r)}{dv / dr}\bigg|_{r=R} \tag{4.2}$$

Then the volumetric flow rate, taking into account the slip Q_s, is defined as

$$Q_S = \int_0^R 2\pi r \cdot v(r)dr = \frac{\pi\left[(d/2)^4 + 4(d/2)^3 \cdot L_s\right]}{8\eta}\frac{\partial p}{\partial z} \tag{4.3}$$

Equation (4.3) is a modified Hagen–Poiseuille equation, taking into account slippage. In the absence of slip, $L_s = 0$ (4.3) coincides with the Hagen–Poiseuille flow Eq (3.5) for the volumetric flow rate without slip Q_P. In [17,18] the parameter enhancement flow ε is introduced. It is defined as the ratio of the calculated volumetric flow rate of slippage to Q_P (calculated using the effective viscosity and the diameter of the CNT). If the measured flux is modeled using Eq (4.3), the degree of enhancement takes the following form

$$\varepsilon = \frac{Q_S}{Q_P} = \left[1 + 8\frac{L_S(d)}{d}\right]\frac{\eta_\infty}{\eta(d)} \tag{4.4}$$

where

$d = 2R$, diameter of CNT

η_∞, viscosity of water

$L_S(d)$, CNT slip length depending on the diameter

$\eta(d)$, the viscosity of water inside CNTs depending on the diameter

-If $\eta(d)$ is found to be equal to η_∞, then the influence of the effect of slip on ε is significant, if $L_S(d) \geq d$

- If $L_S(d) < d$ and $\eta(d) = \eta_\infty$, then there will be no significant difference compared to the Hagen–Poiseuille flow with no slip.

Table 4 summarizes the experimentally measured values of the enhancement water flow. Enhancement flow factor and the length of the slip were calculated using the equations given above.

Figure 38 shows the change in viscosity of the water and the length of the slip in diameter. As shown in the figure, the dependence of slip length to the diameter of the nanotube is well described by the empirical relation.

$$L_S(d) = L_{S,\infty} + \frac{C}{d^3}$$ (4.5)

where

$L_{S,\infty}$ = 30 nm; slip length on a plane sheet of graphene,
C; const

TABLE 4   Experimentally measured values of the enhancement water flow

Nanosystems	Diameter (nm)	Enhancement, ε	Slip length, L_S, (nm)
Carbon nanotubes	300–500	1	0
	44	22–34	113–177
carbon nanotubes	7	10^4–10^5	3900–6800
	1,6	560–9600	140–1400

Figure 39 shows dependence of the enhancement of the flow rate ε on the diameter for all seven CNTs.

The following are three important features in the results:

-   First, the enhancement of the flow decreases with increasing diameter of the CNT.
-   Second, with increasing diameter, the value tends to the theoretical value of Eqs. (4.4) and (4.5) with a slip $L_{g\infty}$ = 30 nm and the effective viscosity $\mu(d) = \mu_\infty$. The dotted line shows the curve of 15% in the second error in the theoretical data of viscosity and slip length.
-   Third, the change ε in diameter of CNTs cannot be explained only by the slip length.

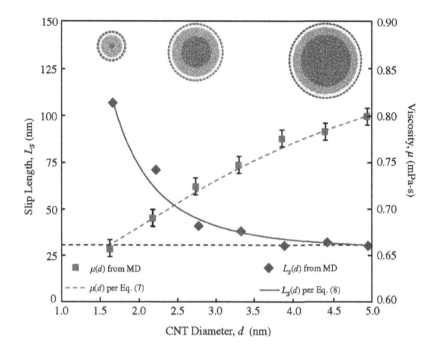

FIGURE 38   Variation of viscosity and slip length with CNT diameter.

To determine the dependence, the molecular modeling method was used by Thomas John A and McGaughey Alan JH [13] with a volumetric flow of water from the pressure gradient along the axis of single-walled nanotube with radii of 1.66, 2.22, 2.77, 3.33, 3.88, 4.44, and 4.99 nm. Snapshot of the water-CNT is shown in Figure 40.

Figure 40 shows the results of calculations to determine the pressure gradient along the axis of the nanotube with the diameter of 2.77 nm and a length of 20 nm. Change of the density of the liquid in the cross-sections was less than 1%.

Figure 41 shows the dependence of the volumetric flow rate from the pressure gradient for all seven CNTs. The flow rate ranged from 3 to 14 m/sec. In the range considered here, the pressure gradient $(0-3).10^9$ atm/m Q (pl/sek $=10^{-15}\,m^3/sek$) is directly proportional to $\partial p/\partial z$. Coordinates of chirality for each CNT are indicated in Figure 41. The linearity of the relations between flow and pressure gradient confirms the validity of calculations of the formula (4.3).

Figure 42 shows the profile of the radial velocity of water particles in the CNT with diameter 2.77 nm. The vertical dotted line at 1.38 nm marked the surface of the CNT. It can be noted that the velocity profile is close to a parabolic shape.

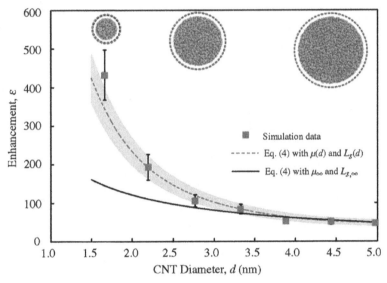

FIGURE 39   Flow enhancement as predicted from MD simulations [13].

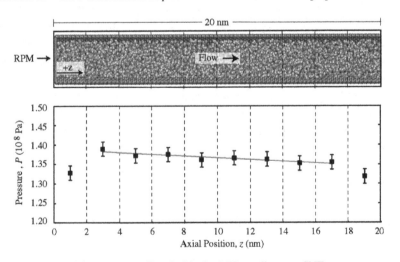

FIGURE 40   Axial pressure gradient inside the 2.77-nm diameter CNT.

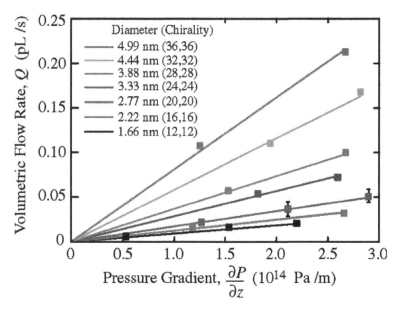

FIGURE 41   Relationship between volumetric liquid flow rate in carbon nanotubes with different diameters and applied pressure gradient.

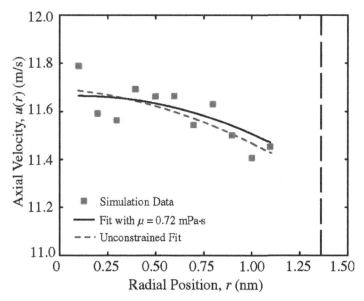

FIGURE 42   Radial velocity profile inside 2.77 nm diameter CNT.

In contrast to previous work [19], the flow of water under a pressure gradient considered in the single-walled nanotubes of "chair" type of smaller radii: 0.83, 0.97, 1.10, 1.25, 1.39, and 1.66 nm.

Figure 43 shows the dependence of the mean flow velocity $\bar{v}$ on the applied pressure gradient $\Delta P / L$ in the long nanotubes is equal to 75 nm at 298 K. A similar picture pattern occurs in the tube with a length of 150 nm.

As one can observe, there is conformance with the Darcy law that the average flow rate for each CNT increases with increasing pressure gradient. For a fixed value of $\Delta p / L$ however, the average flow rate does not increase monotonically with increasing diameter of the CNTs, as derived from Poiseuille equation. Instead, when at the same pressure gradient, the average speed in a CNT decreases with the radius of 0.83 nm to a CNT with the radius of 1.10 nm, similar to the CNTs 1.10 and 1.25 nm, then the speed increases in a CNT with the radius of 1.25 nm to a CNT of 1.66 nm.

The nonlinearity between $\bar{v}$ and $\Delta P / L$ are the result of inertia losses (i.e., insignificant losses) in the two boundaries of the CNT. Inertial losses depend on the speed and are caused by a sudden expansion, abbreviations, and other obstructions in the flow.

Molecular modeling in the work of John A. Thomas, Alan JH McGaughey, and Ottoleo Kuter-Arnebeck [20] shows that Eq (4.1) (Poiseuille parabola) describes the velocity profile of liquid in a nanotube when the diameter of a flow is 5–10 times more than the diameter of the molecule (≈ 0.17 nm for water).

In Figure 44 [20], the effect of slip on the velocity profile at the boundary of radius R of the pipe and fluid is shown. When $L_s = 0$ the fluid velocity at the wall vanishes and the maximum speed (on the tube axis) exceeds flow speed twice.

The figure shows the velocity profiles for Poiseuille flow without slip ($L_s = 0$) and with slippage $L_s = 2R$. The flow rate is normalized to the speed corresponding to the flow without slip. Thick vertical lines indicate the location of the pipe wall. The thick vertical lines indicate the location of the tube wall.

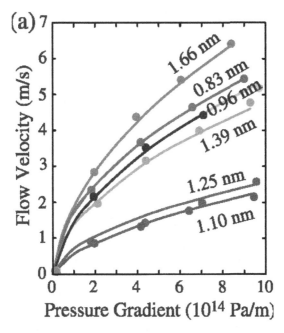

FIGURE 43 Relationship between average flow velocity and applied pressure gradient for the 75 nm long CNTs.

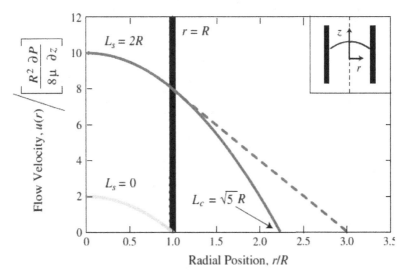

FIGURE 44 No-slip Poiseuille flow and slip Poiseuille flow through a tube.

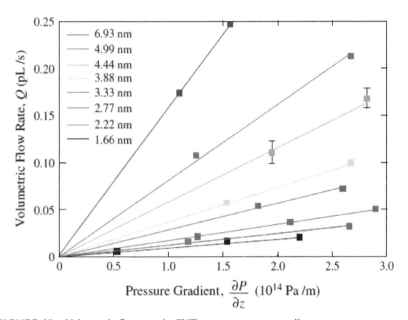

FIGURE 45   Volumetric flow rate in CNTs versus pressure gradient.

Velocity of the liquid on a solid surface can also be quantified by the co-efficient of slip, L_c. The coefficient of slippage is a difference between the radial position in which the velocity profile would be zero. Slip coefficient is equal to $L_c = \sqrt{R^2 + 2RL_s} = \sqrt{5}R$.

For linear velocity profiles (e.g., Couette flow), the length of the slip and slip rate are equal. These values are different for the Poiseuille flow.

Figure 45 shows dependence of the volumetric flow rate Q from the pressure gradient $\partial p / \partial z$ in long nanotubes with diameters between 1.66 nm and 6.93 nm. Pressure gradient Q is proportional to $\partial p / \partial z$. As in the Poiseuille flow, volumetric flow rate increases monotonically with the diameter of CNT at a fixed pressure gradient. Magnitudes of calculation error for all the dependencies are similar to the error for the CNT diameter 4.44 nm (marked in Figure 45).

Researchers [21] considered the steady flow of incompressible fluids in a channel width $2h$ under action of the force of gravity ρg or pressure gradient $\partial p / \partial y$ (which is described by the Navier–Stokes equations). The velocity profile has a parabolic form

$$U_y(z) = \frac{\rho g}{2\eta} \cdot \left[(\delta + h)^2 - z^2 \right]$$

where

δ is the length of the slip, which is equal to the distance from the wall to the point at which the velocity extrapolates to zero.

7.4.1 MODELING IN NANOHYDROMECANICS

We take into consideration the mean free path of gas under normal conditions is 65_{HM} and the distance between the particles—$3{,}3_{HM}$:

$$\frac{4}{3}\pi\delta^3 n = 1, \qquad \delta_{cmaнd} = 3{,}3 \; нм, \qquad \lambda = \frac{1}{\sqrt{2}\pi d^2 n}, \qquad \lambda_{cmaнd} = 65 \; нм,$$

where

n, the concentration of molecules in the air

Knudsen number $= \lambda/L$

L, the characteristic size

d, diameter microtubes

Let us consider the fluid flow through the nanotube. Molecules of a substance in a liquid state are very close to each other (Figure 46).

The molecules of most liquids have a diameter of about 0.1 nm. Each molecule of the fluid is "squeezed" on all sides by neighboring molecules and for a period of time $\left(10^{-10} - 10^{-13} \text{s}\right)$ fluctuates around certain equilibrium position, which itself from time to time is shifted in distance commensuration with the size of molecules or the average distance between molecules, l_{cp}:

$$l_{cp} \approx \sqrt[3]{\frac{1}{n_0}} = \sqrt[3]{\frac{\mu}{N_A \rho}},$$

where

n_0, number of molecules per unit volume of fluid

N_A, Avogadro's number

ρ, fluid density

μ, molar mass

Estimates show that one cubic of nanowater contains about 50 molecules. This gives a basis to describe the mass transfer of liquid in a nanotube-based continuum model. However, the specifics of the complexes, consisting of a finite number of molecules, should be borne in mind. These complexes, called clusters in literatures, are intermediately located between the bulk matter and individual particles (atoms or molecules). The fact of heterogeneity of water is now experimentally established [22].

There are groups of molecules in liquid as "microcrystals" containing tens or hundreds of molecules. Each microcrystal maintains solid form. These groups of molecules or "clusters" exist for a short period of time, then break up and are recreated. Besides, they are constantly moving so that each molecule does not belong at all times to the same group of molecules, or "cluster."

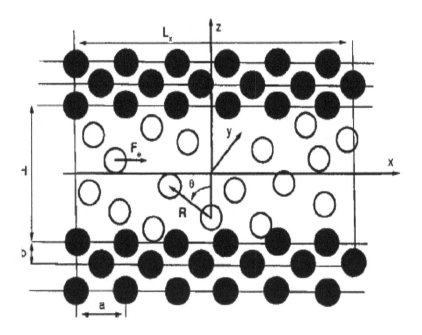

FIGURE 46   The fluid flow through the nanotube.

Modeling predicts that gas molecules bounce off the perfectly smooth inner walls of the nanotubes as billiard balls, and water molecules slide over them without stopping. Possible cause of unusually rapid flow of water is maybe due to the small-diameter nanotube molecules move on them in an order, rarely colliding with each other. This "organized" move is much faster than usual chaotic flow. However, the mechanism of flow of water and gas through the nanotubes is not very clear and only further experiments and calculations can help understand it.

The model of mass transfer of liquid in a nanotube proposed in this research is based on the availability of nanoscale crystalline clusters in it [23].

A similar concept was developed in Ref. [24], in which the model of structured flow of fluid through the nanotube is considered. It is shown that the flow character in the nanotube depends on the relation between the equilibrium crystallite size and the diameter of the nanotube.

Figure 47 shows the results of calculations by the molecular dynamics of fluid flow in the nanotube in a plane (a) and three-dimensional state (b). The figure shows the ordered regions of the liquid.

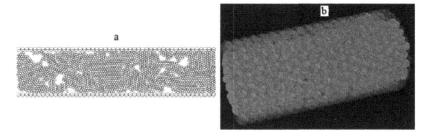

FIGURE 47   The results of calculations of fluid flow in the nanotube.

The typical size of crystallite is 1–2 nm, that is, compared, for example, with a diameter of silica nanotubes of different composition and structure [24].

The flow model proposed in the present work is based on the presence of "quasisolid" phase in the central part of the nanotube and liquid layer, nonautonomous phases [25].

Consideration of such a structure that is formed when fluid flows through the nanotube is also justified by the aforementioned results of the experimental studies and molecular modeling.

When considering the fluid flow with such structure through the nanotube, we will take into account the aspect ratio of "quasisolid" phase and the diameter of the nanotube. Then, the character of the flow is stable and the liquid phase can be regarded as a continuous medium with viscosity η.

Let us establish the relationship between the volumetric flow rate of liquid Q flowing from a liquid layer of the nanotube length l, the radius R, and the pressure drop $\Delta p / l$, $\Delta p = p - p_0$, where

p_0 is the initial pressure in the tube (Figure 48).

Let R_0 be a radius of the tube from the "quasisolid" phase and

v is the velocity of fluid flow through the nanotube.

Structural regime of fluid flow (Figure 49) implies the existence of the continuous laminar layer of liquid (the liquid layer in the nanotube) along the walls of a pipe. In the central part of a pipe a core of the flow is observed, where the fluid moves, keeping its former structure, that is, as a solid ("quasi-solid" phase in the nanotube). The velocity slip is shown in Figure 49(a) through v_0.

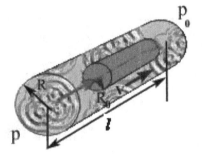

FIGURE 48   Flow through liquid layer of the nanotube.

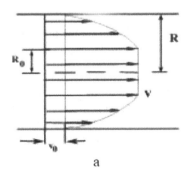
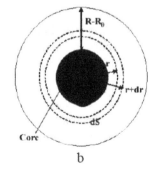

FIGURE 49   Structure of the flow in the nanotube.

Let us find the velocity profile $v(r)$ in a liquid interlayer $R_0 \leq r \leq R$ of the nanotube. We select a cylinder with radius r and length l in the interlayer, located symmetrically to the center line of the pipe (see Figure 49(b)).

At the steady flow, the sum of all forces acting on all the volumes of fluid with effective viscosity η is zero.

The following forces are applied on the chosen cylinder: the pressure force and viscous friction force affect the side of the cylinder with radius r, calculated by the Newton formula.

Thus,

$$(p - p_0)\pi r^2 = -\eta \frac{dv}{dr} 2\pi r l \qquad (4.6)$$

Integrating expression (4.6) between r and R with the boundary conditions $r = R : v = v_0$, we obtained a formula to calculate the velocity of the liquid layers located at a distance r from the axis of the tube:

$$v(r) = (p - p_0)\frac{R^2 - r^2}{4\eta l} + v_0 \qquad (4.7)$$

Maximum speed $v_{\text{я}}$ has the core of the nanotube $0 \leq r \leq R_0$ and is equal to

$$v_{\text{я}} = (p - p_0)\frac{R^2 - R_0^2}{4\eta l} + v_0 \qquad (4.8)$$

Such structure of the liquid flow through nanotubes considering the slip is similar to a behavior of viscoplastic liquids in the tubes. For viscoplastic fluids, a characteristic feature is that they are to achieve a certain critical internal shear stresses τ_0 and behave like solids. Meanwhile, when internal stress exceeds a critical value, the fluids begin to move as normal fluid. In the work of Popov I. Yu et al. [24], the liquid behaves in a similar manner in the nanotube. A critical pressure drop is also needed to start the flow of liquid in a nanotube.

Structural regime of fluid flow requires existence of continuous laminar layer of liquid along the walls of pipe. In the central part of the pipe flow with core radius R_{00} is observed, where the fluid moves, keeping its former structure (i.e., as a solid).

The velocity distribution over the pipe section with radius R of laminar layer of viscoplastic fluid is expressed as follows:

$$v(r) = \frac{\Delta p}{4\eta l}\left(R^2 - r^2\right) - \frac{\tau_0}{\eta}\left(R - r\right) \tag{4.9}$$

The speed of flow core in $0 \leq r \leq R_{00}$ is equal

$$v_{\text{я}} = \frac{\Delta p}{4\eta l}\left(R^2 - R_{00}^2\right) - \frac{\tau_0}{\eta}\left(R - R_{00}\right) \tag{4.10}$$

Let us calculate the flow or quantity of fluid flowing through the nanotube cross-section S at a time unit. The liquid flow dQ for the inhomogeneous velocity field flowing from the cylindrical layer of thickness dr, which is located at a distance r from the tube axis is determined from the relation:

$$dQ = v(r)dS = v(r)2\pi r dr \tag{4.11}$$

where dS is the area of the cross-section of cylindrical layer (between the dotted lines in Figure 49).

Substituting Eq (4.7) into Eq (4.11), integrating over the radius of all sections from R_0 to R, and taking into account that the fluid flow through the core, the flow is determined from the relationship $Q_{\text{я}} = \pi R_0^2 v_{\text{я}}$. Then, we get the formula for the flow of liquid from the nanotube:

$$Q = \pi R^2 v_0 + Q_P\left[1 - \left(\frac{R_0}{R}\right)^4\right] \tag{4.12}$$

If $(R_0/R)^4 < 1$ (no nucleus) and $v_0\Delta p R^2/8l\eta << 1$ (no slip), then Eq (4.12) coincides with Poiseuille formula (3.5). When $R_0 \approx R$ (no of a viscous liquid interlayer in the nanotube), the flow rate Q is equal to volumetric flow $Q \approx \pi R^2 v_0$ of fluid for a uniform field of velocity (full slip).

Accordingly, flow rate of the viscoplastic fluid flowing with a velocity (4.7) is equal to

$$Q = -\frac{\pi R^3 \tau_0}{3\eta}\left[1 - \left(\frac{R_{00}}{R}\right)^3\right] + Q_P\left[1 - \left(\frac{R_{00}}{R}\right)^4\right] \tag{4.13}$$

Comparing Eqs (4.7)–(4.10), (4.12), and (4.13), we can see that the structure of the flow of the liquid through the nanotubes, considering the slippage, is similar to that of the flow of viscoplastic fluid in a pipe of the same radius R.

Given that the size of the central core flow of viscoplastic fluid (radius R_{00}) is defined by

$$R_{00} = \frac{2\tau_0 l}{\Delta p} \tag{4.14}$$

for viscoplastic fluid flow, we obtain Buckingham formula:

$$Q = Q_P\left[1 + \frac{1}{3}\left(\frac{2l\tau_0}{R\Delta p}\right)^4 - \frac{4}{3}\left(\frac{2l\tau_0}{R\Delta p}\right)\right] \tag{4.15}$$

We will establish a conformity of the pipe that implements the flow of a viscoplastic fluid with a fluid-filled nanotube, the same size and with the same pressure drop. We say that an effective internal critical shear stress τ_{0ef} of viscoplastic fluid flows, which ensures the coincidence rate with the flow of fluid in the nanotube. Then from Eq (4.15), we obtain equation of fourth order to determine τ_{0ef}:

$$\left(\frac{2l\tau_{0ef}}{R\Delta p}\right)^4 - 4\left(\frac{2l\tau_{0ef}}{R\Delta p}\right) = A, \ A = 3(\varepsilon - 1), \ \varepsilon = Q/Q_P \tag{4.16}$$

The solution of (4.16) can be found, for example, using the iteration method of Newton:

$$\overline{\tau}_{0ef\,n} = \overline{\tau}_{0ef\,n-1} - \frac{\overline{\tau}_{0ef\,n-1}^4 - 4\overline{\tau}_{0ef\,n-1} - A}{4\overline{\tau}_{0ef\,n-1}^3 - 4}, \ \overline{\tau}_{0ef} = \frac{2l\tau_{0ef}}{R\Delta p} \tag{4.17}$$

The first component in Eq (4.12) represents the contribution to the fluid flow due to the slippage, and it becomes clear that the slippage significantly enhances the flow rate in the nanotube, when $l\eta v_0 >\approx \Delta p R^2$.

This result is consistent with experimental and theoretical results of [17,26–29], which shows that water flow in nanochannels can be much higher than under the same conditions, but for the liquid continuum.

_ In the absence of slippage $\varepsilon = 1$, the Eq (4.16) has a trivial solution $\tau_{0ef} = 0$.

7.4.2   THE RESULTS OF THE CALCULATIONS

Let us determine the dependence of the effective critical inner shear stress τ_{0ef} on the radius of the nanotubes, by taking necessary values for calculations $\varepsilon = Q / Q_P$.[13] The results of calculations at $\Delta p / l = 2.1 \cdot 10^{14}$ Pa/m are summarized in the table below:

R , м	τ_{0ef} (Па)	$\varepsilon = Q / Q_P$
$0.83 \cdot 10^{-9}$	498498	350
$1.11 \cdot 10^{-9}$	577500	200
$1.385 \cdot 10^{-9}$	632599	114
$1.665 \cdot 10^{-9}$	699300	84
$1,94.10^{-9}$	782208	68
$2.22 \cdot 10^{-9}$	855477	57
$2.495 \cdot 10^{-9}$	932631	50

Calculations show that the value of effective internal shear stress depends on the size of the nanotube.

Figure 50 shows the dependence τ_{0ef} on the nanotube radius.

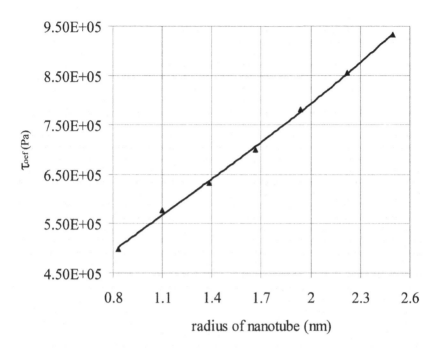

FIGURE 50 Dependence of the effective inner shear stress from the radius of the nanotube.

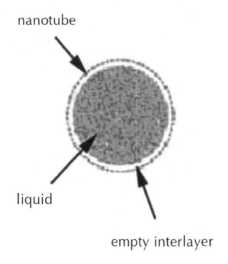

FIGURE 51 The structure of the flow.

The dependence $\tau_{0ef}(R)$ is almost linear. Within the range of the considered nanotube sizes τ_{0ef} has a relatively low value, which indicates the smoothness of the surface of CNTs (Figure 51).

7.4.3   THE FLOW OF FLUID WITH AN EMPTY INTERLAYER

Previous studies [30,31] were analyzed in the aforementioned analysis of the structure of liquid flow in CNTs. The results of the calculations of the cited works (Figures 42 and 43) showed that during the flow of the liquid particles, an empty layer between the fluid and the nanotube is formed. The area near the walls of the CNT $R_* \leq r \leq R$ becomes inaccessible for the molecules of the liquid due to van der Waals, repulsion forces of the heterogeneous particles of the carbon and water (Figure 2). Moreover, according to previous results [30,31], the thicknesses of the layers $R_* \leq r \leq R$ regardless of radii of the nanotubes are practically identical: $R_* / R \approx 0.88$.

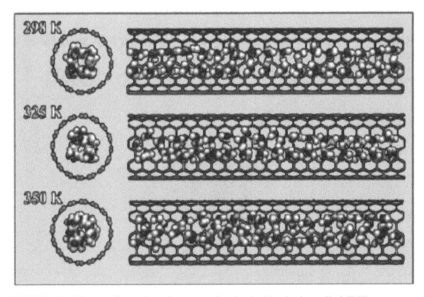

FIGURE 52   The configuration of water molecules inside single-walled CNTs.

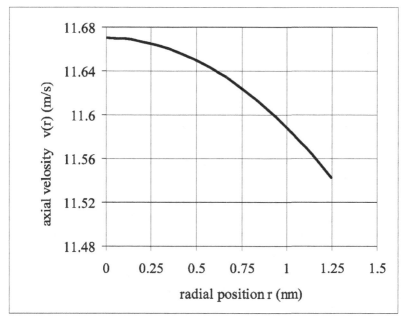

FIGURE 53 Profile of the radial velocity of water in CNTs.

A similar result was obtained in [32], which is an image (Figure 53) of the configuration of water molecules inside (8,8) single-walled CNTs at different temperatures: 298, 325, and 350°K.

Integrating expression (4.6) between r and R_* at the boundary conditions $r = R_* : v = v_*$, we obtain a formula to calculate the velocity of the liquid layers located at a distance r from the axis of the tube:

$$v(r) = v_* + \frac{2Q_p}{\pi R^2}\left[\left(\frac{R_*}{R}\right)^2 - \left(\frac{r}{R}\right)^2\right] \quad (4.18)$$

Substituting Eq. (4.18) in Eq. (4.11) and integrating over the radius of all sections from 0 to R_*, we obtain a formula for the flow of liquid from the nanotube:

$$Q = Q_P\left[v_* \frac{8l\eta R_*^2}{\Delta p R^4} + \left(\frac{R_*}{R}\right)^4\right] \quad (4.19)$$

or

$$\varepsilon = \frac{8l\eta v_*}{\Delta p R^2}\left(\frac{R_*}{R}\right)^2 + \left(\frac{R_*}{R}\right)^4$$

from which we can determine the unknown v_*:

$$v_* = \frac{Q_P}{\pi R^2}\left[\varepsilon - \left(\frac{R_*}{R}\right)^4\right]\left(\frac{R}{R_*}\right)^2$$

(4.20)

Figure 52 shows the profile of the radial velocity of water particles in a CNT with a diameter of 2.77 nm, calculated using the formula (4.18) at $Q_p = 4.75\cdot10^{-19}\,\text{n/m}^2$, $\varepsilon = 114$. The velocity at the border v_* is equal to 11.55 m/s. It is seen that the velocity profile is similar to a parabolic shape, and at the same time agrees with the calculations [13] obtained by using the model of molecular dynamics.

The calculations suggest the following conclusions. Flow of liquid in a nanotube was investigated using synthesis of the methods of the continuum theory and molecular dynamics. Two models are considered. The first is based on the fact that fluid in the nanotube behaves like a viscoplastic. A method of calculating the value of limiting shear stress is proposed, which was dependent on the nanotube radius. A simplified model agrees quite well with the results of the molecular simulations of fluid flow in CNTs. The second model assumes the existence of an empty interlayer between the liquid molecules and wall of the nanotube. This formulation of the task is based on the results of experimental works known from the literature. The velocity profile of fluid flowing in the nanotube is practically identical to the profile determined using molecular modeling.

As seen from the results of the calculations, the velocity value varies slightly along the radius of the nanotube. Such a velocity distribution of the fluid particles can be explained by the lack of friction between the molecules of the liquid and the wall due to the presence of an empty layer. This leads to an easy slippage of the liquid and, consequently, anomalous increase in flow compared to the Poiseuille flow.

7.5 NANOPHENOMENON IN OIL PRODUCTION

Oil-saturated layers are porous materials with different pore sizes, pore chan-nels, and composition of rocks that define the features of interaction forma-tion and injected fluids with the rock. Taking the above-mentioned features into account, we can conclude that the displacement of oil from oil fields in production wells is not a mechanical process of substitution of oil displacing it with water, but a complex physical–chemical process in which the decisive role is played by the phenomenon of ion exchange between reservoir and in-jected fluids with the rock, that is, nanoscale phenomena.

The mechanism of displacement of oil in the reservoir and its recovery is largely determined by the molecular-surface processes occurring at phase interfaces (the rock-forming minerals saturate the reservoir fluid- and gas-displacing agents). Therefore, the problem of wettability is one of the major problems in oil and gas fields of nanoscience.

As the clay is an ultrasystem, a huge amount of research on regulation of the position of clay minerals in porous media with good reason can be attrib-uted to nanoscience. To this the study of gas hydrates should also be included, a number of processes regulating the properties of the pumped oil- and gas-trapped water, water–oil preparation.

Filtering oil in reservoirs at a depth of 13 km of hard rock is determined by the hydrodynamics of the Darcy law. Of the Navier–Stokes equation:

$$\rho\left(\frac{v}{t}+(\nabla\cdot v)v\right)=-\nabla p+\mu v, \tag{5.1}$$

At very low Reynolds numbers, it follows that we can neglect the inertial forces and simplify viscous friction:

$$-\nabla p-\mu\frac{v}{d^2}=0, \tag{5.2}$$

where d characteristic pore size. Equation (5.2) yields Darcy's law

$$v=-\frac{K}{\mu}\nabla p, \tag{5.3}$$

where

$K=d^2$, permeability of the medium

μ viscosity of the fluid

Typical permeability values range from 5 to 500 mD. The permeability of coarse-grained sandstone is $10^{-8} - 10^{-9}$ sm^2 and the permeability of dense sandstone around 10^{-2} sm^2. Medium with the permeability 1-D passes of fluid flow with a viscosity 1 sP at a pressure gradient laym/sm.

During filtration oil gets filled and moves in the pores with a size of $_1$ mkm. To displace oil from the reservoir, one should have a medium with density and viscosity of oil and the size of 1mkm.

It is necessary to conduct experimental and theoretical studies of possible ways to obtain microbubbles and nanoscale environments, to make the calculations and estimates of energy costs upon receipt of such media in different ways and their application to problems of oil production.

7.5.1 PETROLEUM COMPOSITION

Chemically, oil is a complex mixture of hydrocarbons (HCs) and carbon compounds. It consists of the following elements: carbon (84–87%), hydrogen (12–14%), oxygen, nitrogen, and sulfur (1–2%). The sulfur content can reach up to 3–5%. Oils can contain the following parts: a HC, asvalto-resinous, porphyrins, sulfur, and ash. Oil has a dissolved gas that is released when it comes to the earth's surface.

The main part of petroleum HCs are different in their composition, structure, and properties, which may be in gaseous, liquid, and solid states. Depending on the structure of the molecules they are classified into three classes—paraffinic, naphthenic, and aromatic. But a considerable proportion of oil is HCs of mixed structure containing structural elements of all three abovementioned classes. The structure of the molecules determines their chemical and physical properties.

Carbon is characterized by its ability to form chains in which the atoms are connected in series with each other. In remaining connections, hydrogen atoms are attached to the carbon. The number of carbon atoms in the molecules of paraffinic HCs exceeds the number of hydrogen atoms twice, with some constant excess in all the molecules equal to 2. In other words, the general formula of this class of HCs is $C_n H_{2n+2}$. Paraffinic HCs are chemically more stable and refer to the limiting HC.

Depending on the number of carbon atoms in the molecule, HCs may be in one of the three states of aggregation. For example, if there are one to four carbon atoms in a molecule (H_4-C_4H_0), the HC is a gas, from 5 to 16 (C_5H_6 – C_6H_3)—a liquid HC, and if more than 16 (C_7H_5 и т.д.)—solid.

Thus, paraffin HCs in oil can be represented by gas, liquid, and solid crystalline substances. They have different effects on the properties of oil: gas reduces viscosity and increases the vapor pressure.

Fluid paraffins dissolve well in oil only at elevated temperatures, forming a homogeneous mixture. Hard paraffins also dissolve well in oil forming the true molecular mixtures. Paraffin HCs (with the exception of ceresin) can be easily crystallized in the form of plates and plate strips.

Naphthenic (tsiklanovae or alicyclic) HCs have cyclic structure (C/C_nH_{2n}); to be exact, they are composed of several groups—CH_2—interconnected in ringed system. Oil contains mainly naphthenes consisting of five or six groups of CH_2. All connections of carbon and hydrogen are saturated, hence the naphthenic oil has stable properties. Compared with paraffin, naphthenes have a higher density and lower vapor pressure and have better solvent power.

Aromatic HCs (arena) are represented by the formula C_nH_n and have less hydrogen. The molecule has a form of a ring with unsaturated carbon connections. The simplest representative of this class of HCs is benzene C_6H_6, which consists of six groups of C and H. For aromatic HCs, a large solubility, higher density, and boiling point are typical.

Asphalt-resinous portion of oil is a substance of dark color, which is partially soluble in gasoline. They have the ability to swell in solvents, and then pass into mixture.

The solubility of asphaltenes in the resin-carbon systems increases with decreasing concentration of light HCs and increasing concentrations of aromatic HCs.

The resin does not dissolve in gasoline and is a polar substance with a relative molecular mass of 500–1200. They contain the bulk of oxygen, sulfur, and nitrogen compounds of oil.

Asphaltic-resinous substances and other polar components are surface-active compounds and natural oil–water emulsion stabilizers.

Special nitrogenous compounds of organic origin are called porphyrins. It is assumed that they were formed from animal hemoglobin and chlorophyll of plants. These compounds are destroyed at temperatures of 200–250°C.

Sulfur is prevalent in petroleum and HC gas and is contained both in the free state and in the form of compounds (hydrogen sulfide, mercaptans).

Ash is the residue that is formed by burning oil. This is a different mineral compound, usually iron, nickel, vanadium, and sometimes sodium.

Properties of oil determine the direction reprocessing and affect the products derived from petroleum, hence there are different types of classification, which reflect the chemical nature of oil and determine possible areas of processing.

For example, in the base of the classification, reflecting the chemical composition is laid the preference content of one or more classes of HCs in the oil: naphthene, paraffin, paraffin–naphthene, paraffin–naphthene–aromatic, naphthene–aromatic, and aromatic HCs are the different types. Thus, all fractions in the paraffin oils contain a significant quantity of alkanes.

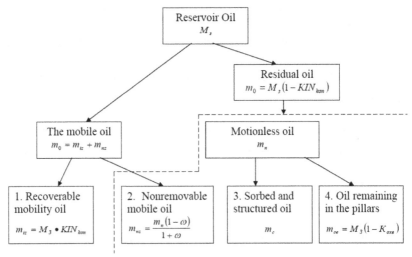

FIGURE 54   The constituents of reservoir oil.

In the paraffin–naphthene–aromatic HCs, all three classes are contained in approximately equal amounts. Naphthene–aromatic oil is characterized mainly by the content of cycloalkanes and arenes, especially in heavy fractions.

Classification is also used by the content of asphaltenes and resins.

In the technical classification, oils are divided into classes according to the sulfur content.

Types: by the output of factions at certain temperatures

Groups: by the potential content of base oils

Species: by content of solid alkanes (of paraffins)

Figure 54 shows the components of the reservoir oil, which have different average integration over the period of development and the entire volume of the reservoir values of physicochemical properties. Here,

M_3, reserves of reservoir oil at reservoir conditions

m_B, mass of water in the reservoir area drained by the end of development, t

ω, watering at the end of development, the proportion of units

m_n, mass of mobile oil, t

$m_{и3}$, mass of extracted oil, t

m_{H3}, mass of the nonremovable movable oil, nonremovable t

m_c, mass of the adsorbed and structured oil t

$m_{Ц}$, mass of oil remaining in pillars, t

The mobile oil (m_n) is part of the reservoir oil moving along layer due to the impact of external influences.

Recoverable mobile oil ($m_{и3}$) is certain part of the mobile oil, which can be extracted from the reservoir as a result of industrial activity with taking into account of economic and technological limitations.

Nonremovable mobile oil (m_{H3}) is part of the mobile oil, which will not be extracted from the reservoir using the technologies as a result of industrial activity on the economic and technological constraints.

The residual oil (m_o) is part of the reservoir oil, located in the reservoir at the end of the displacement.

Motionless oil (m_H) is part of the reservoir oil, remaining motionless in the reservoir due to external influence.

Sorbed and structured oil (m_c) is part of motionless oil, retained near the surface of the collector by the intermolecular interaction.

Oil remaining in pillars ($m_{Ц}$) is part of motionless oil, is not involved in the process of drainage.

Based on the proposed separation of produced oil into separate components, the average integral value of the physicochemical properties of produced oil $\overline{X}_3$ for counting of reserves must be found from the expression:

$$\overline{X}_{_3} = \frac{X_{_{u3}} \cdot m_{_{u3}} + X_{_{H3}} \cdot m_{_{H3}} + X_{_c} \cdot m_{_c} + X_{_{II}} \cdot m_{_{II}}}{M_{_3}} ; \qquad (5.4)$$

$$M_{_3} = m_{_{u3}} + m_{_{H3}} + m_{_c} + m_{_{II}} \qquad (5.5)$$

or

$$X_{_3} = X_{_{u3}} d_{_{u3}} + X_{_{H3}} d_{_{H3}} + X_{_c} d_{_c} + X_{_{II}} d_{_{II}} \qquad (5.6)$$

where

$X_{u3}, X_{H3}, X_c, X_{II}$: the mean of the integral value of the corresponding component of reservoir oil

$d_{u3}, d_{H3}, d_c, d_{II}$: the mass fraction of the corresponding component of reservoir oil $d_i = m_i / M_3$

where

m_i, the mass of i component of reservoir oil,

M_3, the mass of reservoir oil (geological reserves of oil reservoir).

Formulas (5.4) and (5.6) allow us to calculate the average integral value of the physicochemical properties of reservoir oil by using a different source (relative or absolute).

The average integral value of the physico-chemical properties of mobile oil X_n for use in the calculation of oil displacement processes must be found by the expression:

$$\overline{X}_{n} = \frac{X_{_{u3}} \cdot m_{_{u3}}}{m_{_n}} + \frac{X_{_{H3}} \cdot m_{_{H3}}}{m_{_n}} ; \qquad (5.7)$$

or

$$\overline{X}_{n} = X_{_{u3}} c_{_{u3}} + X_{_{H3}} c_{_{H3}} \qquad (5.8)$$

where

$X_{u3} c_{u3} = m_{H3} / m_n$: property value and the mass fraction of component of recovered part of mobile oil, respectively;

$X_{H3} c_{H3} = m_{H3} / m_n$: Property value and the mass fraction of component of nonremovable part of mobile oil, respectively.

In practice, there are a number of contradictions in the calculation methods of geological reserves of HCs and the calculation of oil recovery processes. For example, an anachronism in the method of counting, which is based on the volume (sealer) ratio of produced oil, because the value of reserves is obtained depending on the conditions of oil, as the magnitude of the volume ratio is dependent on the parameters of technology of oil preparation (the number of stages of oil separation and the temperature and pressure conditions).

The density is determined for each formation zone in the case of inhomogeneity of the properties of reservoir oil in the layer. In this case, during the separation of the extracted products on the commodity oil and associated gas, only their masses will be dependent on the properties of reservoir oil and the parameters of ground preparation.

7.5.2 NANOBUBBLES OF GAS, OIL IN THE PORE CHANNELS AND WATER

The average diameter of the channels of the porous medium d is easy to estimate using the well-known relationship:

$$d = \frac{4m}{S}$$
(5.9)

where

m, porosity of the medium, fraction of a unit;

S, specific surface (surface per a volume).

Clay oil gas motherboard thicknesses (OGMT) usually characterized by a porosity of 10–20 percent, and their specific surface of not less than 10^8 m^{-1}. Consequently, the diameters of the pore channels of clay OGMT does not exceed an average of 3 nm.

The minimum diameter of a gas bubble in a water medium can be calculated from the following assumptions.

The gas pressure p_g in a bubble is caused by the action of two components: the deposit pressure $p_{пл}$ and the pressure of surface tension, p_σ.

$$p_g = p_{пл} + p_\sigma$$
(5.10)

or in accordance with the law of Laplace

$$P_g = p_{nɪ} + \frac{2\sigma}{R} \tag{5.11}$$

where

σ, is the coefficient of the surface tension of the liquid, in which formed a gas bubble; R is the bubble radius [33,34].

Gas bubbles can be considered an ensemble of HC molecules with a mass equal to or less than the buoyant force acting on it in the liquid, which gives a formal record

$$NMg = \rho Wg \tag{5.12}$$

where

N, number of molecules in the ensemble

M, mass of each molecule

g, acceleration of free fall

ρ, density the liquid

W, volume of the bubble

The molecular energy of the ensemble, keeping a bottle from collapse (i.e., from the dissolution of gas), is defined by the van der Waals, forces, in which force of intermolecular interaction can be ignored, as the gas has virtually no internal pressure. In this case, the magnitude of this energy per unit volume of a gas bubble is equal to the pressure p_0 in the bubble:

$$p_0 = \frac{NkT}{W - W_0} \tag{5.13}$$

where

T, absolute temperature of the gas $^0 K$

k, Boltzmann's constant

$W_0 = Nw_0$

$w_0 = 12w$, volume per one molecule of gas at the critical temperature and pressure

w, the real volume of a gas molecule

From Eq. (5.13) after simple transformations, taking into account expression for w_0,

$$N = \frac{p_0 W}{kT + p_0 w_0} \tag{5.14}$$

which after substitution in Eq. (5.12) gives

$$p_0 = \frac{kT}{M/\rho - w_0} \tag{5.15}$$

The condition of the bubbles floating necessarily implies the existence of the phase boundary, that is, surface bounding the volume of the bubble. A necessary and sufficient condition for such a boundary is the equality of pressures in gas and liquid phases:

$$p = -p_0 \tag{5.16}$$

Substituting the values of pressures from Eqs. (5.11) to (5.15) gives the desired diameter of the bubble in the form of

$$d_n = 2R = \frac{4\sigma\left(Vp_0 - M/\rho\right)}{kT + p_{nn}\left(w_0 - M/\rho\right)} \tag{5.17}$$

Calculation shows that the diameter of the molecule for the simplest HC gas methane is equal to 0.38 nm, $w_0 = 0{,}345 \cdot 1^{-27} \, m^3$.

If this is taken into account, $M = 27{,}2 \cdot 10^{-27} \, kg$, $\rho \sim 10^3 \, kg/m^3$, $k = 1{,}38 \cdot 10^{-23} \, J/°K$ $T = 360°K$, $\sigma = 62{,}3 \cdot 10^{-3} \, n/m$, then from formula (5.17) it is easy to determine the value of the minimum diameter of the gas bubble.

Under hydrostatic pressure of petroleum $p_{nn} = 20$ Mpa, it is 7 nm. This is more than twice the diameter of the pore channels. Consequently, in the cramped conditions of the pore space of OGMT formation of a gas bubble is impossible, because the capillary pressure in the ducts with a diameter of 3 nm to more than five times higher than the pressure p_0 at a given temperature. This violates the necessary condition for the existence of a gas bubble in a liquid (5.16) and prevents phase separation, as the external pressure leads to a collapse of gas bubbles, that is, to its "dissolution" of the pore fluid, if it occurs at all.

It is clear that the required number of methane molecules to form a bubble is equal to

$$N = \frac{W_n}{w_0} = \frac{179{,}6 \cdot 10^{-27}}{0{,}345 \cdot 10^{-27}} = 520$$

where

W_n, volume of the bubble with diameter of 7 nm

The minimum diameter of a bubble of oil in water is calculated based on approximately similar considerations.

A bubble of oil in an aqueous medium can be considered an ensemble of molecules, which are able to form an interface between the liquid phases. The above (average) radius of the volume of such an ensemble can be obtained from Eq. (5.11):

$$R = \frac{2\sigma}{p_g - p_{nл}}$$

(5.18)

where

σ, border tension coefficient in the water–oil, equal to approximately $45 \cdot 10^{-3}$ n/m ; $p_g = p\,p_{nл} + p,$
where

p, additional molecular pressure in the equation of van der Waals forces, which for a liquid at conditions of incompressibility, is possible to express from the same equation of van der Waals forces:

$$p = -\left(\frac{kT}{w} + p_{nл} \right)$$

(5.19)

where

w, the real volume of one molecule of the liquid, and "minus" sign due to the fact that the pressure force directed toward the center of the bubble of oil.

Then from Eqs (5.18) and (5.19), finally

$$d_n = 2R = \frac{4\sigma w}{kT + wp_{nл}}$$

(5.20)

The resulting formula determines the minimum value of the diameter of the bubble of oil in a water medium. In this case, the real volume of the HC molecule with one carbon atommethane, $w \sim 3{,}5 \times 10^{-28} M^3$.

The total mass of HCs that are brought by foliation flow per unit area of contact OGMT rock reservoir is defined as follows:

$$G = \int_0^t V_G dt \qquad (5.21)$$

where

V_G, velocity of HC-generating organic materials (OM) from OGMT

t, time

G, the productivity of generation

The formation of HCs from the dispersed organic material (DOM) can be regarded as an elimination process in which the initial component is the reactive part of the DOM, and the final product (HC molecules). Then, the reaction velocity of elimination is derived by the well-known formula:

$$V = \varepsilon\left(\Gamma - x\right) = \varepsilon \Gamma e^{-\varepsilon t} \qquad (5.22)$$

as $x = \Gamma\left(1 - e^{-\varepsilon t}\right)$.

In such a formulation

V, the velocity of formation HC from DOM

ε, integrated constant of reaction velocity

Γ, the initial concentration of the reaction capable of part of the DOM in the breed

x, the mass of HCs that has developed in time t

ε, is usually determined experimentally from the velocity of formation of HC out from the given DOM at different temperatures.

Equation (5.22) is valid for open systems in which the products of the reaction easily draw off the center of the reaction. In conditions of porous medium of natural OGMT with a limited pore volume, it is necessary to introduce a factor considering the difficulty of derivatives removal (reaction products) into the formula (5.22).

As noted above, gas (or oil) cannot exist in the pore channels of clay medium in the form of separate phase; the elimination reaction can continue only until the capacity of the pore space is depleted relative to the HC material.

The magnitude of this capacity is determined by the solubility of HC in the pore water under specified pressure and temperature. The limiting value of the concentration of derivatives in the layer place express the maximum capacity of the reaction volume. The velocity of reaction considering this case is limited by the introduction of (5.22) factor $\alpha = 1 - C/C_0$, where C is the current concentration of HC in the pore water and C_0 is its maximum value.

Thus, the velocity of generation of HC per unit area of the roof of OGMT is implemented in OGMT by thickness h, but all products of this generation pass through the surface of contact generating thickness with the collector. Therefore, the amount of generation is convenient to normalize the surface area of the roof OGMT.

$$\text{VG} = \varepsilon g \Gamma \ (1 - C/C_0) e^{-\varepsilon t} \tag{5.23}$$

where

h, thickness of OGMT

The magnitude of the current concentration of HCs in the flow of pore water, squeezed out from OGMT, by definition, is equal to

$$C = \frac{x}{W} = \frac{h\Gamma}{W}\left(1 - e^{-\varepsilon t}\right) \tag{5.24}$$

where

W, the volume of pore water, released as a result of compaction OGMT and/or passed through it over time t

x, the mass of the produced HC substances

Substituting Eqs. (5.23) and (5.24) into (5.21),

$$G = \varepsilon h \Gamma \int_0^t \left[1 - \frac{h\Gamma\left(1 - e^{-\varepsilon t}\right)}{C_0 W}\right] e^{-\varepsilon t} dt \tag{5.25}$$

which after integration and simple transformations gives

$$G = h\Gamma\left(1 - e^{-\varepsilon t}\right)\left[1 - \frac{h\Gamma}{2WC_0}\left(1 - e^{-\varepsilon t}\right)\right] \tag{5.26}$$

According to Eq. (5.23), the obtained formula remains valid until $C < C_0$. When $C = C_0$ the HC generation ceases. If the current HC concentration exceeds a specified limit (i.e., HC products of degradation are beginning to stand out in a separate phase, forming gas bubbles or droplets of oil), then the formula loses physical meaning.

Formula (5.26) is valid during the primary migration of HCs and loses its physical meaning in the transition to the process of secondary migration. From this, it follows that the inequality $G > 0$ is always valid, $G = 0$ at $t = 0$ and when $t \to \infty$, G tends to $G_0 = h\Gamma$, which is quite realistic.

As $W = w_0 + W_\Phi$

where

W_Φ, volume of foliation, which flow through unit area of OGMT by thickness h, and w_0, volume of pore fluid contained in a block of OGMT with a single base and height h, value $h\Gamma/2C_0W$ is always less than 1. However, w_0 is the volume in which the reaction of degradation of DOM occurs. It is equal to the volume of voids of OGMT, including the volume occupied by the DOM. This can be written as $W_0 = hn + h\Gamma/\rho$, which in turn suggests that the inequality

$$\frac{h\Gamma}{2C_0W} = \frac{h\Gamma}{2C_0\left(hn + h\Gamma/\rho + W_\phi\right)} < 1$$

It should be noted that the density of DOM does not exceed $2/\text{t/m}^3$, and is often significantly less; $h/n > h\Gamma$ in most cases is of practical interest, and $2C_0$ at a depth of petroleum is always more than 1 kg/m^3.

As formulated in the task, the movement of pore fluid in the clay thickness in the inviolate natural environment under the principles of geofluid dynamics of slow flow is realized in the foliation regime. It is absolutely clear that the alien pore water molecules of HC derivatives will be forced to fall on the axis of the pore channels with greater frequency than the water molecules. This is due to the fact that near the walls of the pore channels, molecules of water are additionally bonded by surface forces of mineral skeleton, whereas in the center of the channel the resultant of these forces is equal to zero [35]. Similar conclusion was obtained by P. A. Diskej [36]. Therefore, the concentration of HCs in foliation flow is greater than C_0. However, the cramped conditions of the pore space in OGMT, as noted, do not allow such molecules to combine into a separate phase. Consequently, they are forced to migrate in a homogeneous unstable mixture with molecules of the pore water.

Based on the above, one can calculate the concentration of HCs in foliation stream, if their concentration in the volume of pore space is equal to C. The ideology of this calculation is fairly obvious.

If the channel pore with radius R and the length l contains N molecules of a solvent such as water, then

$$R^2 l = N \cdot \frac{4}{3}\pi r^3$$

where

r is the radius of molecules of the solvent, the total number of which is n_t at the fixed moment of time are located on the channel axis(i.e., there are always vacancies). Then $l = 2rnt$, which after substituting into the initial equation gives

$$n_t = \frac{2}{3} N \frac{r^2}{R^2}$$

In other words, the number of molecules that fall at the same time on the axis of the pore channel, in R^2/r^2 time smaller than two-thirds of their total number N in the channel. And because the frequency of contact of the HC molecules with the axis of the pore channel is in prior compared with the water molecules (Figure 55); their concentration in this part of the pore space is equal to

$$C_\phi = C\frac{N}{n_t} = \frac{3}{2}C\frac{R^2}{R^2} \tag{5.27}$$

The resulting formula characterizes the concentration of HC substances in foliation flow (Eq. (5.27) holds for the reaction volumes commensurate with the volume occupied by the molecules of derivatives, which is characteristic of the pore space of clay OGMT.

A capillary pressure force directed against the forces of buoyancy occurs at the HC cluster in contact with manifold overlapping tight layer, and therefore

$$K_{np} \leq \frac{n_n}{2}\left[\frac{gH}{\sigma}(\rho_0 - \rho) + \sqrt{\frac{n}{2K}}\right]^2 \tag{5.28}$$

where

$K_{\Pi p}$ and n_{n}, coefficients of permeability and porosity of the proposed tires, respectively

K and n, coefficients of permeability and porosity of the manifold respectively

σ, border tension between the JC-phase and the reservoir water

H, depth of reservoir

g, acceleration of free fall

ρ_{0}, density of reservoir water

ρ, density of the HC phase

From the known Arrhenius equation, the coefficient of the reaction velocity of degradation is equal to

$$\varepsilon = Ae^{-E/RT} \tag{5.29}$$

where
  E, the activation energy of degradation

  R, universal gas constant

  T, absolute temperature of the system

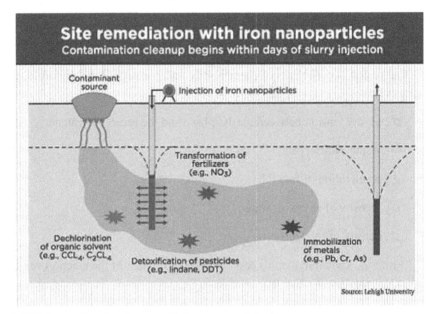

FIGURE 55   Cleaning the area with iron nanoparticles.

7.5.3   PROPERTIES OF ALUMINUM

Aluminum powder reacts with water at 50°C with an allocation of hydrogen. It reacts vigorously in exothermic reactions with oxygen-containing liquids with halogenated organic compounds and other oxidants.

Aluminum is relatively nontoxic, relatively inexpensive, widely distributed in nature, and produced in large quantities in industries by electrolysis.

Aluminum in the form of nanopowder has lowered the ability to reaction at room temperature due to the presence of dense oxide–hydroxide shell, which is an electric double layer.

7.5.4   THE USE OF ALUMINUM POWDER IN OIL PRODUCTION

Aluminum powder is used in the metallurgical industry in aluminothermy, as alloying additives for the manufacture of semifinished products by pressing and sintering. Very strong components (gears, bushings, etc.) are obtained with this method. Powders are also used in chemistry for the preparation of compounds of aluminum as a catalyst (e.g., production of ethylene and acetone).

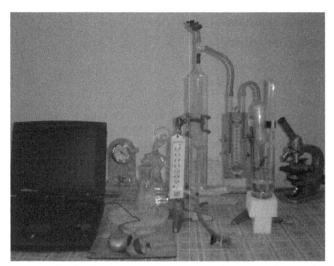

FIGURE 56   The picture of a set in which the pressure is measured.

Considering high reactivity of aluminum, especially in powder form, it is used in explosives and solid fuel for rockets, using its property to quickly ignite.

In the oil industry, nanoparticles of aluminum were used for the separation of oil–water emulsion.

Figure 56 shows a snapshot of the device, where osmotic pressure was measured. The pressure difference Δp was measured with a liquid manometer—U-shaped tube. Manometer was connected to a closed receptacle, which was partially vacuumed to avoid the effects of atmospheric pressure Δp.

The use of aluminum powder in injection wells for water injection leads to a significant increase in reservoir pressure, creating the effect of hydrobreak and facilitates the efficient displacement of oil.

KEYWORDS

- **Carbon nanotubes**
- **Hydrocarbons**
- **Molecular synamics**
- **Nanohydromechanics**
- **Slippage**
- **van der Waals, forces**
- **Viscoplastic fluid**

BIBLIOGRAPHY

1. Olgica, B.; et al. Separation of 100-kilobase DNA molecules in 10 seconds, *Anal. Chem.* **2001,** *73(24),* 6053–6056.

2. Chou, C.-F.; et al. Sorting biomolecules with microdevices, <http://www.ncbi.nlm. nih.gov/   pubmed/10634473##>Electrophoresis<http://www.ncbi.nlm.nih.gov/ pubmed/10634473##>. **2003,** *21(1),* 81–90.

3. Uchic1 Michael, D.; Dimiduk1 Dennis, M.; Florando Jeffrey, N.; Nix William, D.; Sample dimensions influence strength and crystal plasticity. *Sci.* August 13, **2004,** *305(5686),* 986–989.

4. Suetin, M. V.; and Vakhrushev, A. V.; Molecular dynamics simulation of adsorption and desorption of methane storage managed nanocapsules. "All-Russian Conference with international participation the Internet" "From nanostructures, nanomaterials and nanotechnologies for nanotechnology," *Izhevsk,* 08/04/2009, 112 p.

5. Yoon, J.; and Ru, C. Q.; and Mioduchowski A. *Compos. Sci. Technol.* **2003,** *63,* 1533.

6. Wang, C. Y.; Ru, C. Q.; and Mioduchowski A. *Phys. Rev. B* **2005,** *72,* 075414.

7. Natsuki, T.; Endo, M.; and Tsuda, H.; *J. Appl. Phys.* **2006,** *99,* 034311.

8. Yoon, J.; Ru, C. Q.; and Mioduchowski A. *J. Appl. Phys.* **2003,** *93,* 4801.

9. Yoon, J.; Ru, C. Q.; and Mioduchowski A. *J. Compos. B* **2004,** *35,* 87.

10. Natsuki, T.; Hayashi, T.; and Endo, M.; *J. Appl. Phys.* **2005,** *97,* 044307.

11. Wang, Q.; and Varadan, V. K.; *Int. J. Solids Struct.* **2006,** *43,* 254.

12. Morten Bo Lindholm Mikkelsen, Simon Eskild Jarlgaard, and Peder Skafte-Pedersen.; Experimental Nanofluidics. Capillary filling of nanochannels. MIC—Department of Micro and Nanotechnology Technical University of Denmark, June 20, **2005.**

13. Thomas, J, A.; and McGaughey, A. J. H.; Reassessing fast water transport through carbon nanotubes. *Nano Lett.* **2008,** *8(9),* 2788–2793.

14. Joseph, P.; and Tabeling, P.; *Phys. Rev. E.* **2005,** *71,* 035303.

15. Barrat, J.-L.; and Chiaruttini, F.; *Mol. Phys.* **2003,** *101,* 1605–1610.

16. Sokhan, V. P.; Nicholson, D.; and Quirke, N. J.; *Chem. Phys.* **2002,** *117,* 8531–8539.

17. Holt, J. K.; Park, H. G.; Wang, Y.; et al. Fast mass transport through sub-2nm carbon nanotubes. *Sci.* **2006,** *312,* 1034–1037.

18. Majumder et al., **1999**
19. Thomas, J. A.; and McGaughey, A. J. H.; Water flow in carbon nanotubes: transition to subcontinuum transport. prl 102, *Phys. Rev. Lett.* **2009,** 184502–1–184502-4.
20. Thomas, J. A.; McGaughey, A. J. H.; and Kuter-Arnebeck, O.; Pressure-driven water flow through carbon nanotubes: Insights from molecular dynamics simulation. *Int. J. Therm. Sci.* **2010,** *49,* 281–289, journal homepage: www.elsevier.com/locate/ i j t s.
21. Whitby, M.; and Quirke, N.; Fluid flow in carbon nanotubes and nanopipes. Chemistry Department, Imperial College, South Kensington, London SW7 2AZ, UK. *Nat. nanotechnol.* www.nature.com/naturenanotechnology, February **2007,** *2,* 87–94.
22. Huang, C.; Wikfeldt, K. T.; Tokushima, T.; et al. The inhogeneous structure of water at ambient conditions. Proceeding of the National Academy of Sciences. http:/www. pnas.org/content/early/2009/08/13/0904743106.abstract.
23. Frenkel, Ya. I.; *UFN,* **1941,** *25(1),* 1–18.
24. Popov, I. Yu.; Chivilikhin, S. A.; and Gusarov, V. V.; Model of the structured liquid through the nanotube: http://rusnanotech09.rusnanoforum.ru/Public/ LargeDocs/theses/rus/poster/04/Chivilikhin.pdf.
25. Gusarov, V. V.; Popov, I. Yu.; Il Nuovo Cimento, D.; **1996,** *18(7),* 799–805.
26. Kalra, A.; Garde, S.; and Hummer, G.; Osmotic water transport through carbon nanotube arrays. *Proc. Natl. Acad. Sci. USA.* **2003,** *100,* 10175–10180.
27. Majumder, M.; Chopra, N.; Andrews, R.; and Hinds, B.; *J. Nat.* **2005,** *438,* 44–44.
28. Skoulidas, A. I.; Ackerman, D. M.; Johnson, J. K.; and Sholl, D. S; *Phys. Rev. Lett.* **2002,** *89,* 185901.
29. Hummer, G.; Rasaiah, J. C.; and Noworyta, J. P.; *Nat.* **2001,** *414,* 188–190.
30. Kotsalis, E. M.; Walther, J. H.; and Koumoutsakos, P.; Multiphase water flow inside carbon nanotubes. *Int. J. Multiphase Flow.* **2004,** *30,* 995–1010.
31. Chen, X.; Qiao, Y.; Zhou, Q.; and Cao, G.; *Mol. Simul.* **2008,** *88,* 371–378.
32. Hongfei, Ye.; Hongwu, Z.; Zhongqiang, Z.; and Yonggang, Z.; Size and temperature effects on the viscosity of water inside carbon nanotubes. *Nanoscale Res. Lett.* **2011,** 6–87, http://www.nanoscalereslett.com/content/6/1/87.
33. Nigmatulin, R. I.; Dynamics of multiphase media. Part I, *M.:* "*Nauka,*" **1987,** 464 p.
34. Nagiyev, F. B.; Nonlinear oscillations of gas bubbles dissolved in the liquid. Izv.AN Az.SSR, ser.f.-tech and math. *Sci.* **1985,** *1,* 136–140.
35. Arie, A. G.; and Slavkin, B. C.; About the mechanism of oil gas saturation of sand lenses. *Oil and Gas Geol.* **1995,** *2.*
36. Diskej, P. A.; Possible primary migration of oil from source rock in oil phase. *Bull. AAPG.* **1975,** *59(2).*
37. "Biomolecules with Microdevices," *Electrophoresis.* **2000,** *21(1),* 81–90.
38. Bortov, V. Yu.; Garanin, D. I.; Georgievski, V. Yu.; et.al. Comparative tests of reforming catalysts by "AKSENS." *Petrochem. Ref.* **2003,** *2,* 10–17.
39. Buyanov, R. A.; and Krivoruchko, O. P.; Development of the theory of crystallization of sparingly soluble metal hydroxides and scientific basis of catalyst preparation of compounds of this class. *Kinet. Catal.* **1976,** *17(3),* 765–775.
40. Chernoivanov, V. I.; Mazalov, Yu. A.; Soloviev, R. Yu.; et al. A method of making the composition. RF Patent on application No. 2003119564 from July 02, **2003.**
41. Chemical Encyclopedia. Ed. Knunyants, I. L.; Moscow: Soviet Encyclopedia; **1990,** *1,* 2.

42. Churaev, N. V.; Ralston, J.; Sergeeva, I. P.; and Sobolev, V. D.; Electrokinetic properties of methylated quartz capillaries. *J. Colloid and Interface Sci.* **2002**, *96,* 265–278.

43. Cottin-Bizonne, C.; Barentin, C.; Charlaix, E.; et al. Dynamics of simple liquids at heterogeneous surfaces: molecular-dynamics simulations and hydrodynamic description. *Eur. Phys. J. E.* **2004**, *15,* 427–438.

44. Cottin-Bizonne, C.; Cross, B.; Steinberger, A.; and Charlaix, E.; Boundary slip on smooth hydrophobic surfaces: intrinsic effects and possible artifacts. *Phys. Rev. Lett.* **2005**, *94,* 056102.

45. Culligan, Patricia J.; Taewan, K.; and Yu, Q.; Nanoscale fluid transport: size and rate effects. *Nano Lett.* **2008**, *8(9),* 2988–2992.

46. Darrigol, O.; Between hydrodynamics and elasticity theory: the first five births of the Navier–Stokes equations. *Arch. Hist. Exact Sci.* **2002**, *56,* 95–150.

47. Ermolenko, N. F.; and Efres, M. D.; Regulation of the Porous Structure of Oxide Adsorbents and Catalysts. Moscow: Nauka; **1991**.

48. Evstrapov, A. A.; The Course of Lectures "Nanotechnology in Environment and Medicine," **2011**, 136 p.

49. Ficini, J.; Lumbroso-Bader, N.; and Depeze, J-K.; Fundamentals of physical chemistry. – *M.: Mir*, **1972**.

50. Friction laws at the nanoscale. *Nat.* Feburary 26, **2009**.

51. Gairik, S; Abhay, P.; Nelson, V.; et al. *J. Chem. Phys.* **2006**, *124,* 144708.

52. Godymchuk, A. Yu.; Ilyin, A. P.; and Astankova, A. P.; Oxidation of aluminum nanopowder in liquid water during heated. Proceedings of the Tomsk Polytechnic University, **2007**, *310(1),* 102–104.

53. Hemanth, G. R.; Will water flow through nanotubes? www.dstuns.iitm. ac.in/teaching-and-presentations/, **2010**.

54. Ilyin, A. P.; Gromov, A. A.; and Yablunovsky, G. V.; About activity of aluminum powders. *Phys. Combust. Explosion.* **2001**, *37(4),* 58–62.

55. Ilyin, A. P.; Godymchuk, A. Yu.; and Tikhonov, D. V.; Threshold phenomena in the oxidation of aluminum nanopowders. Physics and chemistry of ultrafine (nano-) systems: Proceeding. VII All-Russian Conference. Moscow: Moscow Engineering Physics Institute Printing; **2005**, 178–179.

56. Jason, K. H.; Hyung, G. P.; Yinmin, W.; et al. Fast mass transport through sub-2-nanometer carbon nanotubes. *Sci.* **2006**, *312,* 1034–1037.

57. Kozlova, E. G.; Emelianov, Yu. I.; Krasiy, B. V.; et al. The new catalysts for reforming of gasoline with an octane rating of 96-98. *Catal. Ind.* **2003**, *6.*

58. Korchagina, Yu. I.; and Chetverikov, O. P.; Methods of assessing the generation of hydrocarbons produce oil. *M.: Nedra*, **1983**.

59. Lauga, E.; Brenner, M. P.; and Stone, H. A.; Microfluidics: The No-Slip Boundary Condition in Handbook of Experimental Fluid Dynamics. New York: Springer, **2006**.

60. Lauga, E.; and Stone, H. A.; Effective slip in pressure-driven Stokes flow. *J. Fluid Mech.* **2003**, *489,* 55–77.

61. Li, T. D.; Gao, J.; Szoszkiewicz, R.; Landman, U.; and Riedo, E.; *Phys. Rev. B.* **2007**, *75,* 115415.

62. Mazalov, Yu. A.; Patent RF No. 2158396. *The Method of Burning Metal Fuels.* **2000**.

63. Mirzadzhanzade, A. Kh.; Maharramov, A. M.; et al. Study the influence of nanoparticles of iron and aluminum in the process of increasing the intensity of gas release and pressure for use in oil production. News of Baku University. *Sci. Ser.* **2005,** *1,* 5–13.

64. Mirzadzhanzade, A. Kh.; Maharramov, A. M.; Nagiyev, F. B.; and Ramazanov, M. A.; Nanotechnology Applications in the Oil Industry. Proceedings of the II-nd Scientific Conference "Nanotechnology-Production 2005," Fryazino, November 30–December 1, **2005,** 47–52.

65. Mirzadzhanzade, A. Kh.; Maharramov, A. M.; Nagiyev, F. B.; On the development of nanotechnology in the oil industry. *"Azerbaijan's Oil Ind.,"* **2005,** *10,* 51–65.

66. Mirzadzhanzade, A. Kh.; Bakhtizin, R. N.; Nagiyev, F. B.; and Mustafayev, A. A.; Nano hydrodynamic effects on the base use of micro embryonic technology. *"Oil and Gas Business,"* **2005,** *3,* 311–315.

67. Mirzadzhanzade, A. Kh.; Shahbazov, E. G.; Shafiev, Sh. Sh.; Nagiyev, F. B.; Osmanov, B. A.; and Mammadzadeh, R. B.; Nanotechnology in the oil and gas production: research, implementation and results. Book of abstracts. Khazarneftgazyatag—2006. International Scientific Conference on October 25–26, **2006,** 47.

68. Nagiyev, F. B.; and Khabeev, N. S.; Dynamics of soluble gas bubbles. Proceedings of the academy of sciences USSA. *Fluid and Gas Mech.* **1985,** *6,* 52–59.

69. Nagiyev, F. B.; and Mustafin, R. Kh.; Using of High Technologges in Oil Production. Intensification of Oil Production with Aid of Nanohydrodynamic Effects Usage. Collection of Thesis International Workshop "Socio-Economic Aspects of the Energy Corridor Linking the Caspian Region with E.U." Baku, Azerbaijan: April, 11–12th, **2007,** 28–37 p.

70. Navier, C. L. M. H.; Memoire sur les lois du mouvement des fluids. *Mémoires Académie des Sciences de l'Institut de France.* **1823,** *1,* 389–440.

71. Neimark, I. E.; The main factors influencing the porous structure of hydroxide and oxide adsorbents. *Colloid J.* **1982,** *4(4),* 780–783.

72. Nigmatulin, R. I.; Fundamentals of mechanics of heterogeneous media. *M.: "Nauka,"* **1978,** 336 p.

73. Ou, J.; Perot, J. B.; and Rothstein, J. P.; Laminar drag reduction in microchannels using ultrahydrophobic surfaces. *Phys. Fluids.* **2004,** *16,* 4635–4643.

74. Ou, J.; and Perot, J. B.; Drag reduction and μ-PIV measurements of the flow past ultrahydrophobic surfaces. *Phys. Fluids.* **2005,** *17,* 103606.

75. Press release on the website of the University of Wisconsin-Madison. Models present new view of nanoscale friction, February 25, **2009.**

76. Proskurovskaya, L. T.; Physical and Chemical Properties of Electroexplosive Ultrafine Aluminum Powders: Dis. Ph.D., Tomsk: **1988,** 155 p.

77. Ramazanova, E. E.; Shabanov, A. L.; and Nagiyev, F. B.; Perspectives of Nanotechnology Method Applications for Intensification Oil-Gas Production. Collection of Thesis International Workshop "Electricity Generation and Emission Trading in South Eastern Europe." Sofia, Bulgaria: September 21, **2007.**

78. Rothstein, J. P.; and McKinley, G. H.; *J. Non-Newtonian Fluid Mech.* **1999,** *86,* 61–88.

79. Semwogerere, D.; Morris, J. F.; and Weeks, E. R.; *J. Fluid Mech.* **2007,** *581,* 437–451.

80. Sorokin, V. S.; Variational method in the theory of convection. *Appl. Math. Mech.* **1953,** *XVII,* 39–48.

81. Stepin, B. D.; and Tsvetkov, A. A.; Inorganic Chemistry. Moscow: Higher School, **1994,** 608 p.

82. Sunyaev, Z. I.; Sunyaev, R. Z.; and Safiyeva, R. Z.; Oil dispersions systems. -*M.: Chem,* **1990.**

83. Chang, T.; Dominoes in carbon nanotubes. *Phys. Rev. Lett.* October 24, **2008,** *101,* 175501.

84. Wei-xian, Z.; Nanoscale iron particles for environmental remediation: an overview. *J. Nanopart. Res.* **2003,** *5,* 323–332.

85. Xi, C.; Guoxin, C.; Aijie, H.; Venkata, K.; Punyamurtula, L. L.; Patricia, J.;

CHAPTER 8

CORRELATION BETWEEN THE CONDUCTIVE PROPERTIES OF POLY(METHYL METHACRYLATE) AND ITS MICROSTRUCTURE

NEKANE GUARROTXENA and MIGUEL MUDARRA

CONTENTS

8.1   INTRODUCTION

The objective of this chapter is to analyze the influence of a set of tacticity-governed microstructures on the conductive properties of poly(methyl methacrylate) (PMMA). Those structures are specially the mmmr and the mmmmrx (x = m or r) that occur necessarily whenever an isotactic sequence breaks off, and also the rrrm-based termini of syndiotactic sequences. The nature and characterization of all the repeating stereosequences related to tacticity have been widely studied in earlier works to which the reader is referred.[1–7] A straight relation between these repeating stereosequences and a list of physical properties including dielectrical relaxations, glass transition, and electrical space charge nature and distribution, as investigated in our laboratory, have been widely conveyed for PVC and PP polymers.[8–20] The frequency of those repeating stereosequences, the type of the likely local conformations in some of them, the length of the associated tactic sequences, either iso- or syndiotactic, and the atactic parts and the pure mrmr moieties were found to be a major driving force for the inter- and intrachain interactions and, subsequently, for the physical properties studied.

Such an important correlation was explained to obey, on the one hand, microstructures (especially when adopting GTTG–TT conformation), and on the other, the chain alignment and the interchain sequential interactions depend on the regularity of the segments separating the former structures.[11,13,15]

The results first enabled us to assess some original property/microstructure relationships and second to shed light on the mechanisms of the physical processes involved in most of the polymer behaviors. These mechanisms are of crucial interest in the field of materials science. With the purpose of extending these concepts to other materials of industrial interest, some courses of action concerning PMMA have been endeavored. The first attempt deals with the space charge behavior of three PMMA samples of quite different content and distribution of the above-mentioned repeating stereosequences, as carefully analyzed using ¹H-NMR spectroscopy. As published recently [21], the straight relation between the nature and amount of space charge and some repeating stereosequences, namely mmrm, rmmr, and rrrm, as disclosed in an earlier work on PVC and PP samples [13,15,17], holds well for PMMA, hence taking the generalization of that relationship is an important step further.

On the basis of the above results, we study herein the conductive properties of the same PMMA samples through the dynamic electrical analysis technique. In reality, according to the ^{1}H-NMR analysis, these polymers present different content of mmrm termini of isotactic sequences of at least one heptad long, located at different intervals of regular chain segments from one sample to another. As a result, mmrm, which proved to be traps of negative space charges [13,15,17] and exhibited favored rotation facilities, could make the conductivity easy although at different degrees for every sample. The same holds, to distinct extents, for the other quoted stereosequences. The study of the electrical conductivity by dynamic electrical analysis will be conducted at the electrical modulus level.[22–24] A sublinear frequency dispersive AC conductivity has been frequently observed in polymers, so that the real part of the conductivity $\sigma'(\omega)$ can be expressed as follows:

$$s'(w) = s_0 + Aw^n \qquad (1)$$

where σ_0 is the DC conductivity, A is a temperature-dependent parameter, and n is a fractional exponent that ranges between 0 and 1 and has been interpreted by means of many-body interactions among charge carriers. This behavior, termed universal dynamic response, has been observed in highly disordered materials such as ionically conducting glasses, polymers, amorphous semiconductors, and also in doped crystalline solids.[25–33] Equation (1) can be derived from the "universal dielectric response function"[34] for the dielectric loss of materials with free hopping carriers, and this derivation allows one to understand the temperature dependence of parameter A.[35]

In the same manner, the coupling degree involved in the local motions, as defined by Ngai [36–37], is expected to change from one sample to another. In reality, it was shown to be higher as the rrrm, connected with syndiotactic sequences, is more frequent [8–10] and depends markedly on the conformation in the case of mmrm, where GTTG–TT conformation is much lesser coupled than GTGTTT conformation.[8,9] Interestingly, this coupling is connected with the inter- and intrachain interactions in that it involves changes in both local free volume and local motion, and it alters the distance between these points and then the length of the regular sequences that are responsible for chain alignments. Therefore, the distinct behavior of the samples studied herein should be able to furnish valuable insight into the role of the above stereosequences in the conductive properties of PMMA. Elucidating this relationship is the purpose of the present work.

8.2 EXPERIMENTAL SECTION

8.2.1 MATERIALS

PMMA samples (*X*, *Y*, and *Z*) provided from commercial source (Atochem) were used in this work. They were purified using tetrahydrofuran as solvent water as precipitating agent, washed in methanol, and dried under vacuum at 40°C for 48 hours. THF was distilled under nitrogen with aluminum lithium hydride to remove peroxides immediately before use.

8.2.2 ¹H-NMR SPECTROSCOPY

The tacticity of the three distinct PMMA samples was measured using 1H-NMR spectroscopy on a Varian Unity-500 spectrometer operating at 499.88 MHz. Spectra were recorded at 50°C on approximately 10 wt % solutions in deuterochloroform. Typical parameters for the proton spectra were 8,000 Hz spectral width, a 1.9 sec pulse repetition rate, 0.5 sec delay time, and 64 scans. The relative peak intensities were measured from the integrated peak areas, which were calculated with an electronic integrator. The content of the different triads were determined by the electronic integration of the α-methyl signals, which appear at (1.3–1.1 ppm), at (1.1–0.9 ppm), and at (0.9–0.7 ppm) corresponding to isotactic (mm), heterotactic (mr), and syndiotactic (rr) triads, respectively.[38]

8.2.3 SIZE EXCLUSION CHROMATOGRAPHY

The molecular weight distribution was measured by size exclusion chroma-tography (SEC) using a chromatographic system equipped with a Waters Model 410 refractive index detector. THF was used as the eluent at a flow rate of 1 cm³/min at 35°C. Styragel packed columns (HR1, HR3, HR4E, and HR5E Waters Division Millipore) were used. PMMA standards (Polymer Laboratories) in a range between 3.53×10^6 and 5.8×10^2 g/mol were used to calibrate the columns. The ¹H-NMR data and the molecular weight values are summarized in Table 1.

8.2.4 CALORIMETRY

The glass transition temperature (T_g) of the three PMMA samples, performed with a differential scanning calorimeter (DSC), were found equal to 115.6°C, 106°C, and 93.5°C for X, Y, and Z, respectively. About 11 mg of PMMA sample was placed in an aluminum pan and put on a hot plate (170°C, N_2) where it was maintained for 10 min to erase previous thermal history; it was then cooled to 50°C at a fast rate (40°C/min) and a heating DSC run from 50 to 170°C at 2°C/min was carried out.

8.2.5 DYNAMIC ELECTRIC ANALYSIS

The real and imaginary parts of the electrical permittivity were measured at several frequencies at isothermal steps of 5°C each. The measurements were carried out using a Novocontrol BDS40 dielectric spectrometer with a Novotherm temperature control system.

The real and imaginary parts of the electric modulus, $M^*(\omega)$, were calculated from the complex permittivity and were fitted to the real and imaginary parts of the electric modulus given by Eq. (6) simultaneously. Eight independent parameters were used in the fitting process: s_0, w_p, e_{00C}, n, e_{00C}, t_{HN}, α, and β. The physical meaning of these parameters is introduced later, where Eq. (6) is derived.

In this work, we have used simulated annealing to carry out the fitting process. This method has been successfully used in the analysis of thermally stimulated depolarization currents [39,40] and dielectric spectroscopy data. [41]

8.3 RESULTS AND DISCUSSION

8.3.1 MICROSTRUCTURE OF THE SAMPLES

The ^{1}H-NMR spectra of samples X, Y, and Z are compared in Figure 1. The resonance assignments indicated for the mm-, mr-, and rr-centered pentads are those taken from literature.[21] The spectra are typical of predominantly syndiotactic PMMA but showing different isotactic contents. The quantitative amount of mm, mr, and rr triads, as measured on the spectra through direct integration of signals, are summarized in Table 1.

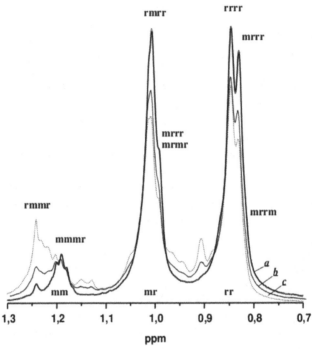

FIGURE 1   500 MHz ^{1}H-NMR spectra of (a) X, (b) Y, and (c) Z PMMA samples measured in CDCl$_3$ at 50°C.

TABLE 1 ^{1}H-NMR data for the PMMA samples: [a]percentage of isotatic (mm), heterotactic (mr), and syndiotactic (rr) triads; [b]percentage of m (P_m = mm + 1/2mr) and rr (P_r = rr + 1/2mr) diads

Sample	M_n	mm[a]	mr[b]	rr[a]	P_m[b]	P_r[b]
X	45,500	8.39	43.16	48.45	29.97	70.03
Y	43,100	15.16	40.86	43.98	35.59	64.41
Z	44,900	20.93	37.92	41.15	38.89	60.11

The fractions of the mm, rr, or mr-centered pentads have been determined from the spectra of Figure 1 by deconvoluting the overall triad signals into the individual (or coupled) pentad signals as indicated in Figure 1. It allows both the deconvolution and the distribution of every experimental triad percentage into the corresponding pentad percentages, through an internal mathematical treatment.

The data so obtained are summarized in Table 2. One way to corroborate their validity is to compare them to the calculated values taking into account the type of repeating sequence statistics, whether Bernoullian or Markovian, of the samples. The extent to which each sample fits into, or departs from, Bernoullian statistics can be determined from the mm, rr, and mr and rm triad content as measured on the spectra (Table 1). That quantity is called the persistent ratio and is defined by $\rho = P(s)P(i)/P(is)$, where $P(s) = rr + 1/2mr$; $P(i) = mm + 1/2mr$, and $P(is) = 1/2mr$. The results obtained are 0.9725, 1.1220, and 1.2709. These values indicate that sample X is clearly Bernoullian ($\rho = 1$), whereas samples Y and Z depart somewhat from this statistics. The same was inferred when considering the conditional probabilities of first-order Markov statistics [42], which is another known criterion to examine the type of repeating sequence statistics.

Therefore, the pentad fractions were calculated considering both samples Y and Z with Markovian statistics and assuming all the samples are Bernoullian. Interestingly, the former values happened to diverge strongly from those measured on the spectra (Table 2), especially in the order of changing from sample X to sample Z. Reversely this order, unlike the absolute values, is quite coincident when considering all the samples as Bernoullian (Table 3). The differences in absolute value are absolutely because samples Y and Z are only majority Bernoullian. Anyway, what is important for the purpose of this chapter are the changes in fraction pentads from one sample to the other, as reflected in Tables 2 and 3.

TABLE 2 The iso-, hetero-, and synditactic pentads values obtained by deconvolution of the triads signals on ^{1}H-NMR spectra of PMMA samples (Figure 1).

		Sample		
Triads	Pentads	X	Y	Z
	rrrr	0.395	0.39	0.39
Rr	mrrr + mrrm	0.089	0.049	0.0245

TABLE 2   *(Continued)*

		Sample		
Triads	Pentads	*X*	*Y*	*Z*
	rmrr + mmrr	0.4197	0.397	0.3070
Mr	mrmr + mmrm	0.0118	0.018	0.072
	rmmr	0.019	0.0406	0.0226
Mm	mmmr + mmmm	0.064	0.111	0.1866

Consequently, the above results are quite valuable to settle the evolution of any repeating stereosequence from one sample to the other. The sequences that were proved to be the major driving force for the physical properties of PVC and PP according to earlier work are (i) the average length of the isotactic and syndiotactic sequences and (ii) the mmr- and the rrm-based local structures, which occur necessarily whenever an isotactic or syndiotactic sequences break off, respectively. Nevertheless, these structures are not active by themselves because it is the occurrence of either one m placement following mmr or one r placement preceding rrm and the length of the $-mm\cdots-$ and $-rrr\cdots-$ sequences connected with them that were identified as a property-determining factor; hence, the most important factors are (a) the fraction of mmr followed by one m placement (−mmrm− structures) and the length of the isotactic sequence preceding mmrm. In fact, the ratio of −mmrm− repeating stereosequences of at least one heptad in length to the same shorter ones was proved to be of major importance; (b) the fraction of rrm preceded by one or more *r* placement, that is, the −rrrm− structures at the end of syndiotactic sequences; (c) the pure heterotactic −mrmr− sequences; and (d) the short atactic moieties such as rmrr, mmrr, mrrm, and rmmr.

TABLE 3   The iso-, hetero-, and syndiotactic pentad values calculated by assuming Bernouillian statistics for PMMA samples

		Sample		
Triads	Pentads	*X*	*Y*	*Z*
rr	Rrrr	0.24	0.17	0.13
	mrrr + mrrm	0.4984	0.3474	0.230

TABLE 3 *(Continued)*

		Sample		
Triads	Pentads	*X*	*Y*	*Z*
	rmrr + mmrr	0.2939	0.295	0.2882
mr	mrmr + mmrm	0.125	0.163	0.191
	rmmr	0.044	0.0525	0.057
mm	mmmr + mmmm	0.0457	0.058	0.0763

The changes of these repeating stereosequences in *X*, *Y*, and *Z* samples can be specified in light of the above-mentioned both calculated and experimental results (Tables 2 and 3). It may be thus concluded as follows:

(i) The average length of isotactic –mmmm– sequences increases from samples *X* to *Z* and so does the content of mmrm sequence and the length of the isotactic sequence preceding it. As a result, the ratio of mmrm stereosequences longer than one heptad to the shorter ones will increase in the order $X < Y < Z$.

(ii) The fraction of –rrrm– structures and the length of the syndiotactic sequence preceding it will decrease in the order $X > Y > Z$.

(iii) As indicated by the individual calculated values, the fraction of pure heterotactic stereosequences, –mrmr····–, hardly changes from one sample to the other. A tiny trend toward decreasing from *X* to *Z* is however observed.

(iv) The short atactic moieties, mrrm, rmrr, and mmrr, decrease in the order $X > Y > Z$, this trend being significant for mrrm only. The rmmr changes in the reverse order.

On the contrary, it has been extensively discussed [1,4] that mmrm can adopt GTTG–TT and GTGTTT conformation, the equilibrium between them being strongly displaced toward the latter conformation. In PMMA, such a displacement should be much enhanced relative to PVC, because of its more hindered rotation facilities. Nevertheless, the occurrence of GTTG–TT conformation will decrease in a similar way to mmrm, that is, $Z > Y > X$.

3.2   DYNAMIC ELECTRIC ANALYSIS MEASUREMENTS

The real and imaginary parts of the electric modulus have been plotted as a function of the frequency for several temperatures above the glass transition in Figure 2. Only sample Z is shown, but similar results can be observed for samples X and Y. In the case of the imaginary part of the modulus, two peaks can be observed in the frequency range studied. The one at higher frequencies is related to molecular motions that result in *alpha* relaxation and the one at lower frequencies is associated with the conductive processes. It is noted that the peak shifts to higher frequencies with temperature.

To study the charge transport process at these temperatures, we have assumed a sublinear frequency dispersive AC conductivity. Power–law dependencies of conductivity, as in the case of Eq. (1), imply a power–law dependence of the form $(j(w))^n$ for the complex conductivity.[29] Therefore, this magnitude can be obtained as follows:

$$s'(w) = s_0 + A(jw)^n + jwe_0 e_{00C} \tag{2}$$

where e_{00C} refers to the permittivity at high frequency. A crossover frequency w_p can be defined, $w_p = \frac{s_0}{A}$, so that Eq. (2) can be rewritten as follows:

$$s = s_0 + s_0 \left(j\frac{w}{w_p} \right)^n + jwe_0 e_{00C} \tag{3}$$

This frequency w_p is associated with the crossover from the power–law dependence observed at high frequency to a frequency-independent DC regime that occurs at low frequencies. Finally, the contribution to the permittivity is

$$e_C = -\frac{js}{e_0 w} \tag{4}$$

The alpha relaxation can be modeled by means of Havriliak–Negami equation

$$e_{HN} = e_{00HN} + \frac{\Delta e}{1 + \left((jwt_{HN})^a \right)^b} \tag{5}$$

where e_{00HN} refers to the permittivity at high frequencies and $\Delta e = e_{00C} - e_{00HN}$ is the relaxation strength. Therefore, the electric modulus over the frequency range considered can be expressed as follows:

$$M^{-} = \left(e_{C}^{-} + e_{HN}^{-}\right)^{-1} \qquad (6)$$

An agreement between experimental data and values calculated after fitting to Eq. (6) can be noted in Figure 2, in which s_0, w_p, e_{00C}, n, e_{00HN}, t_{HN}, α, and β were treated as parameters to be fitted.

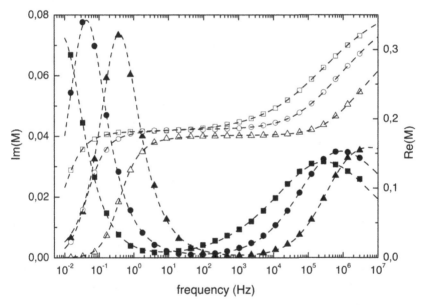

FIGURE 2 Real and imaginary parts of the electric modulus (M) at several temperatures for Z samples. The symbols correspond to experimental values. Lines correspond to values calculated after fitting processes. Im(M): ■ 125°C; ● 140°C; ▲ 160°C. Re(M): □ 125°C; ○ 140°C; △ 160°C.

An Arrhenius plot of the values calculated for the DC conductivity σ_0 is as shown in Figure 3. It can be noted that the conductivity is thermally activated in all cases. The values of the preexponential factor σ_{0o} and the activation energy E_a in Table 4 indicate that there is a correlation between the content of sequences mmrm and the DC conductivity. Indeed, the conductivity is favored by mmrm structure, that is, as the local free volume and rotational motions are enhanced (see NMR analysis, tables, and results reported). In fact, their content increases from X to Z, so making the mmrm-based stereosequences to be closer to the observed favored conductivity. The fact that the sequences mmrm equal to or greater than one heptad act as traps for negative carriers, resulting in well-defined space charge profiles observed by thermal step, supports this hypothesis.[21]

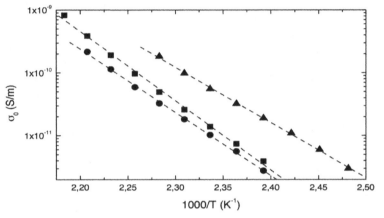

FIGURE 3   Arrhenius plot of DC conductivity σ_0 calculated by fitting Eq. (6) to data. Samples: ■ X; ● Y; ▲ Z.

The temperature dependence of the parameter n is shown in Figure 4. This parameter characterizes the power–law conduction regime, which is associated with the slowing down of the relaxation process in the frequency domain as a result of the cooperative effects, in the same way as the KWW function does in the time domain.

TABLE 4   Preexponential factor σ_{00} and activation energy E_a resulting from fitting DC conductivity σ_0 to Arrhenius law

Sample	σ_{00} (S/m)	E_a (eV)
X	8.54×10^{14}	2.19
Y	3.35×10^{12}	1.99
Z	2.09×10^{10}	1.75

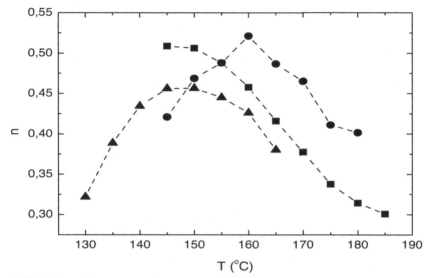

FIGURE 4   Sublinear power–law exponent n vs temperature resulting from fitting Eq. (6) to data. Samples: ■ X; ● Y; ▲ Z.

An important connection between these two approaches stems from the coupling model of Ngai and Kannert [43]. This model predicts power–law conductivity associated with the KWW relaxation function:

$$F(T) = \exp\left[-\left(\frac{t}{t^*}\right)^b \right] \tag{7}$$

where t^* is the relaxation time of KWW function. This power–law conductivity can be obtained as follow:

$$s_\downarrow KWW(w) =^* \exp\left(-E_\downarrow a / kT\right) w^\uparrow (1-b) \tag{8}$$

Therefore, if any other contribution is sufficiently smaller than that of $sKWW$, then the conductivity of the material may be described by $sKWW(w)$.[43] The sublinear dispersive AC conductivity observed in PMMA can be associated with a KWW relaxation mechanism with $\beta = 1 - n$, where n is the power–law exponent determined from $\sigma(\omega)$.

The stretched exponential parameter β may represent an index of correlation of ionic motion, as the stretched exponential relaxation time has been associated with a slowing of the relaxation process that results from correlated ion hopping, so that one would expect β to be close to 0 for strongly correlated systems and close to 1 for random Debye-like hops. As shown in Figure 4, the power–law exponents in all samples have maxima, at approximately the same temperature (145°C) in the case of samples X and Z, and at 160°C in the case of sample Y. Maxima range is between 0.45 and 0.52, indicating that the process does not reach a strongly correlated hopping regime in the temperature range considered.

In all cases, correlation in ion hopping increases initially with temperature, which can be associated with the increase in the carrier concentration, which results in a smaller mean distance and stronger interactions between them. After the maximum, correlation decreases, what we associate with the effect of the increase of local motions of chain segments, which decreases correlated motions of the carriers.

The above results are in agreement with those obtained when studying the space charge behavior of the same samples [21], thereby suggesting the role of rrrm- and especially the mmrm-based repeating stereosequences in the conductive properties. The local free volume and rotational motion are different in the latter, and they could both relate to the nature and frequency of the carriers. In particular, the mmrm, contrary to the rrrm, can adopt the GTGTTT or the GTTG–TT conformations of different free volume and rotational motion. The conformational equilibrium GTGTTT $\rightleftarrows$ GTTG–TT strongly lies to the left side and is necessarily governed by the temperature. It was demonstrated to work toward the right side at temperatures higher than 120°C [44], what may explain the occurrence of maxima in Figure 4. By comparing the behavioral properties of samples of different contents of mmrm- and rrrm-based stereosequences, those processes were argued to be accompanied by changes in parameter n and, consequently, in local or sequential chain correlations. [8–21] The changes in conductive properties obtained herein clearly suggest the same type of correlation. The apparent deviation of sample Y in Figure 4 is presumably due to the offsetting effects of rrrm- and mmrm-based sequences, which are both of intermediate magnitudes between X and Z samples.

Figure 5 shows the correlation between Maxwell relaxation time

$$t_M = \frac{e_0 e_r}{s_0} \tag{9}$$

which characterizes the charge carrier relaxation process and the characteristic time

$$t_p = \frac{2p}{w_p} \tag{10}$$

which determines the transition from DC regime to AC regime. In the case of low temperatures, it seems that both t_M and t_p take the same value (in Figure 5(a) dashed line indicates $t_M = t_p$). At higher temperatures, both t_M and t_p decrease with temperature, as expected, but Maxwell time, associated with carrier mobility is higher than the time that characterizes the regime transition (Maxwell time decreases at a lower rate). This indicates that carrier relaxation time is larger than the characteristic time of the processes involved in AC regime. This may be the reason for the decrease in the carrier motion correlation with the temperature: time that determines carrier mobility is too high to allow carriers to follow processes associated to AC regime at high temperatures.

In the case of the maximum values of n in each case, it can be seen that samples X and Y reach similar values, which are higher than in the case of sample Z. This result agrees with the explanation given above for DC conductivity. There is also a correlation between the content of mmrm-based sequences and the parameter n, and it relates both to the overall content of the latter sequences and to the conformational equilibrium GTGTTT $\rightleftarrows$ GTTG–TT between the conformations that are likely in them. These structures condition the presence of regions with larger free volume and enhanced local rotational motions in regularly spaced locations of the chain, thereby influencing the chain correlations [21] and then the conductive properties.

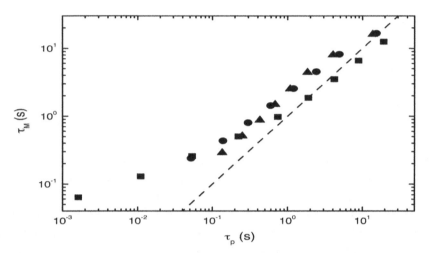

FIGURE 5 Correlation between Maxwell relaxation time (t_M) and the characteristic time (t_p) that determines the transition from DC regime to AC regime. Samples: ■ X; ● Y; ▲ Z. Dashed line is a guide to the eye to indicate $t_p = t_M$.

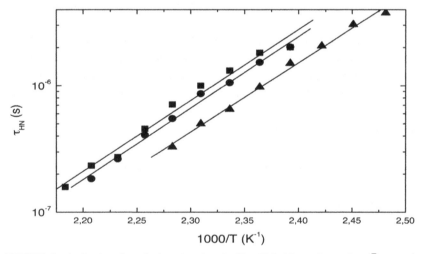

FIGURE 6 Arrhenius plot of relaxation time in Havriliak–Negami equation (τ_{HN} t_{HN}) resulting from fitting Eq. (6) to data. Samples: ■ X; ● Y; ▲ Z.

The relaxation time in Havriliak–Negami equation has been represented in an Arrhenius plot (Figure 6). Havriliak–Negami equation has been used to model the relaxation observed at the high-frequency range observed in our measurements (Figure 2). This relaxation is associated with the cooperative motions of chain segments, which are allowed above the glass transition. The temperature range of the experiments is somewhat above the glass transition temperature of each of the samples (T_g values can be seen in the experimental section). At these temperatures, there are no volume restrictions and it can be noted that kinetics obeys Arrhenius law. The values of the preexponential factors and activation energies corresponding to the samples used are summarized in Table 5.

TABLE 5 Preexponential factor τ_{HNO} and activation energy E_a resulting from fitting relaxation time τ_{HN} to Arrhenius law

Sample	τ_{HNO} (S)	E_a (eV)
X	9.44×10^{-20}	1.11
Y	6.67×10^{-20}	1.12
Z	1.11×10^{-19}	1.08

As can be noted, the correlation between both magnitudes with the GTTG–TT content is quite apparent, even if sample Y exhibits some deviation because of the above-mentioned offsetting effects of rrrm and mmrm sequences relative to X and Z samples. Indeed, the predominance of either rrrm or mmrm stereosequences is much more marked in samples X or Z, respectively.

8.4 CONCLUSIONS

The influence of microstructure on the conductive properties of PMMA has been studied at high temperatures. The electrical properties have been studied at the electrical modulus level, assuming a dispersive conductivity to explain conduction processes and Havriliak–Negami equation to evaluate dipolar contribution with a good agreement between experimental data and the model.

Conductivity is thermally activated at the temperature range considered, and AC regime can be associated with a correlated ion hopping of carrier process, which does not reach a strongly correlated regime. We associate the initial increase in the correlation with the increase in carrier concentration.

Conductivity is concluded to be favored by mmrm-based stereosequences longer than one heptad. According to prior work [21], this exhibit enhanced local free volume and rotational motion, what would involve significant changes in carrier correlation and then in conductivity behavior.

For the lowest temperature values in the range considered, charge relaxation time is quite similar to the times that characterize the transition from DC to AC regime. In the higher temperature range, free carriers relaxation time is greater than the regime transition time, which may also contribute to the decrease in hopping correlation, as carriers cannot follow processes associated with AC regime, which has lower characteristic times.

Kinetics of chain motions at these temperatures follows Arrhenius law, as there are no volume restrictions and the presence of regions with larger free volume and enhanced local rotational motions in regularly spaced locations are associated with the difference in activation energy.

ACKNOWLEDGMENT

MM thanks the Agència de Gestió d'Ajuts Universitaris i de Recerca de la Generalitat de Catalunya for their financial support (2009SGR1168).

KEYWORDS

- PMMA
- Microstructure
- Tacticity
- Conductivity
- Space charge

REFERENCES

1. Guarrotxena, N.; Schue, F.; Collet, A.; and Millán, J.; *J. Polym. Int.* **2003**, *52*, 420–428.
2. Guarrotxena, N.; Martínez, G.; and Millán, J.; *J. Polym. Sci. Polym. Chem.* **1996**, *34*, 2387–2397.
3. Guarrotxena, N.; Martínez, G.; Millán, J.; *J. Polym. Sci. Polym. Chem.* **1996**, *34*, 2563–2574.
4. Guarrotxena, N.; Martínez, G.; and Millán, J.; *Eur. Polym. J.* **1997**, *33*, 1473–1479.
5. Guarrotxena, N.; Martínez, G.; and Millán, J.; *Polym.* **1999**, *40*, 629–636.

6. Martínez, G.; García, C.; Guarrotxena, N.; and Millán, J.; *Polym.* **1999**, *40*, 1507–1514.

7. Guarrotxena, N.; Martínez, G.; and Millán, J.; *Acta Polym.* **1999**, *50*, 180–186.

8. Guarrotxena, N.; Del Val, J.J.; and Millán, J.; *Polym. Bull.* **2001**, *47*, 105–111.

9. Guarrotxena, N.; Del Val, J. J.; Elicegui, A.; and Millán, J.; *J. Polym. Sci. Polym. Phys.* **2004**, *42*, 2337–2347.

10. Del Val, J. J.; Colmenero, J.; Martinez, G.; and Millán, J.; *J. Polym. Sci. Polym. Phys.* **1994**, *32*, 871–880.

11. Guarrotxena, N.; Martinez, G.; and Millán, J.; *Polym.* **1997**, *38*, 1857–1864.

12. Guarrotxena, N.; Martinez, G.; and Millán, J.; *Polym.* **2000**, *41*, 3331–3336.

13. Guarrotxena, N.; Vella, N.; Toureille, A.; and Millán, J.; *Macromol. Chem. Phys.* **1997**, *198*, 457–469.

14. Guarrotxena, N.; Millán, J.; Vella, N.; and Toureille, A.; *Polym.* **1997**, *38*, 4253–4259.

15. Guarrotxena, N.; Vella, N.; Toureille, A.; and Millán, J.; *Polym.* **1998**, *39*, 3273–3277.

16. Guarrotxena, N.; Contreras, J.; Toureille, A.; and Millán, J.; *Polym.* **1999**, *40*, 2639–2648.

17. Guarrotxena, N; Toureille, A.; and Millán, J.; *Macromol. Chem. Phys.* **1998**, *199*, 81–86.

18. Guarrotxena, N.; and Millán, J.; *Polym. Bull.* **1997**, *39*, 639–646.

19. Guarrotxena, N.; Millán, J.; Sessler, G.; and Hess, G.; *Macromol. Rapid Commun.* **2000**, *21*, 691–696.

20. Guarrotxena, N.; Contreras, J.; Martinez, G.; and Millán, J.; *Polym. Bull.* **1998**, *41*, 355–362.

21. Guarrotxena, N.; Retes, J.; Agnel, S.; and Toureille, A.; *J. Polym. Sci. Polym. Phys.* **2009**, *47*, 633–639.

22. Moynihan, C. T.; *Solid State Ionics.* **1998**, *105*, 175–183.

23. Pissis, P.; and Kyritsis, A.; *Solid State Ionics.* **1997**, *97*, 105–113.

24. Macdonald, J. R.; *Solid State Ionics.* **2002,** *150*, 263–279.

25. Jonscher, A. K.; Dielectric Relaxation in Solids. London: Chelsea Dielectric Press; **1983**.

26. Jonscher, A. K.; Universal Relaxation Law. London: Chelsea Dielectric Press; **1996**.

27. Sidebottom, D. L.; Green, P. F.; and Brow, R. K.; *Phys. Rev. Lett.* **1995**, *74*, 5068–5071.

28. Sidebottom, D. L.; Green, P. F.; and Brow, R. K.; *Phys. Rev. B.* **1997**, *56*, 170–177.

29. León, C.; Lucía, M. L.; and Santamaría, J.; *Phys. Rev. B.* **1997**, *55*, 882–887.

30. León, C.; Lucía, M. L.; Santamaría, J.; and Sánchez–Quesada, F.; *Phys. Rev. B.* **1998**, *57*, 41–44.

31. Kalgaonkar, R. A.; Jog, J. P.; *Polym. Int.* **2008**, *57*, 114–123.

32. Sengwa, R. J.; Choudhary, S.; and Sankhla, S.; *Polym. Int.* **2009**, *58*, 781–789.

33. Mahmoud, W. E.; Al–Ghamdi, A. A.; *Polym. Int.* **2010**, *59*, 1282–1288.

34. Jonscher, A. K.; *Phys. Thin Films.* **1980**, *11*, 222–232.

35. Almond, D. P.; West, A. R.; and Grant, R. J.; *Solid Stat. Commun.* **1982**, *44*, 1277–1280.

36. Ngai, K. L.; *Comments Solid State Phys.* **1979**, *9*, 127–140.

37. Ngai, K.; *Comments of Solid State Phys.* **1979**, *9*, 141–155.

38. Bovey, F. A.; High Resolution NMR of Macromolecules. New York-London: Academic Press; **1972**.

39. Laredo, E.; Suarez, N.; Bello, A.; de Gáscue, B. R.; Gómez, M. A.; and Fatou, J. M. G.; *Polym.* **1999,** *40,* 6405–6416.

40. Grimau, M.; Lared, E.; Bello, A.; and Suarez, N.; *J. Polym. Sci. Polym. Phys.* **1997,** *35,* 2483–2493.

41. Bello, A.; Laredo, E.; and Grimau, M.; *Phys. Rev. B.* **1999,** *60,* 12764–12774.

42. Hatada, K.; and Kitayma, T.; NMR Spectroscopy of Polymers, Chapter 3, Springer-Verlag; **2004.**

43. Sidebottom, D. L.; Green, P. F.; and Brow, R. K.; *J. Non-Cryst. Solids.* **1995,** *183,* 151–160.

44. Koenig, J. L.; and Antoon, M. K.; *J. Polym. Sci. Polym. Phys.* **1977,** *15,* 1379–1395.

CHAPTER 9

DEVELOPMENT OF ELECTROHYDRODYNAMICS AND THERMO-ELECTRO-HYDRO DYNAMICS PRINCIPLES FOR ELECTROSPUN FIBER REINFORCED THROUGH *IN SITU* NANOINTERFACE FORMATION

A. K. HAGHI and A. POURHASHEMI

CONTENTS

9.1 INTRODUCTION

Electrospinning is one of the few known methods of creating nanoscale fibers. Nanoscale fibers are attractive materials for their superior surface area-to-volume ratio, making it ideal for important applications. As illustrated in Figure 1, a high-voltage electrode is placed in contact with the polymer solution contained in a pipette or syringe-like vessel with a capillary tip. This electrode provides a source of charge. The ground electrode of the high-voltage source is attached to collector plate, which serves as a target for the electrostatically driven polymer fluid stream. The potential difference between the capillary tip and the ground is typically on the order of 10–30 kV.[1–3]

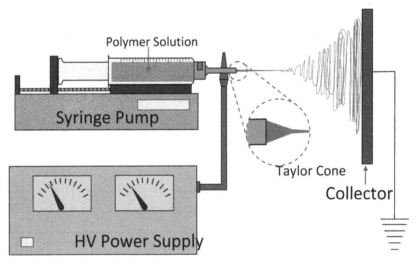

FIGURE 1   A typical electrospinning configuration.

At a sufficiently high potential difference, the electrostatic stresses overcome the surface tension of the Taylor cone. Then, a stream of polymer fluid is ejected and is propelled toward the grounded target. As the ejected stream forms a filament that traverses the distance from the Taylor cone at the capillary tip to the grounded target, the solvent component is lost by evaporation processes and the remaining polymer solidifies into a coherent filament. The motion of the filament is straight for a relatively short distance and then becomes erratic due to an electric field-induced bending instability. The result of this dynamic process is a nonwoven filament mat that collects on the grounded target.[4–5]

The advantages of electrospinning over the other methods of nanofiber fabrication are its ease of manufacture and simple fabrication tools required. And the disadvantages of electrospinning are the lack of control of the final product with respect to fiber diameter, uniformity, and morphology. The inherent instability of the process makes its lack of repeatability problematic. Although much research has been done in the process itself, its wide-scale adoption has been inhibited by a lack of predictive control on the fiber properties. By developing an accurate computational model, enhanced process control and the production of fibers with desired properties can be attained. Electrospinning is an example of an electrohydrodynamic (EHD) phenomenon. In EHDs, charges induce fluid motion within an electric field. During the process, the transport and distribution of these charges generate stresses that result in the movement of the fluid. The leaky dielectric EHD model is an appropriate model to use because the model of the fluid's electrical properties as a poorly conducting liquid is comparable to the behavior of most polymer solutions, the most commonly used type of fluid in electrospinning.[6–7] But it does not take thermal effect into account. For a polymer with high molten temperature, thermal factor is critical for the process. Hence, a rigorous thermo-electro-hydrodynamics description of electrospinning is needed for better understanding of the process.[8] This study establishes a mathematical model to explore the physics behind electrospinning.

9.2   EHD MODEL FOR ELECTROSPINNING PROCESS

In 1969, Melcher and Taylor [7] investigated a branch of fluid mechanics named electrohydrodynamics, which involves the effects of moving media on electric fields for the first time after William Gilbert's experiments in the seventeenth century. Next, in 1997, Savile [6] reported this phenomenon in the book titled *Electrohydrodynamics: the Taylor–Melcher leaky dielectric model* in detail. Applications of EHDs abound: spraying, dispersion of one liquid into another, coalescence, inkjet printing, boiling, augmentation of heat and mass transfer, fluidized bed stabilization, pumping, and polymer dispersion are but a few.

The leaky dielectric model consists of the Stokes equations to describe fluid motion and an expression for the conservation of current employing an ohmic conductivity. Electromechanical coupling occurs only on fluid–fluid boundaries where the charge, carried to the interface by conduction, produces electric stresses different from those present in perfect dielectrics or perfect conductors. With perfect conductors or dielectrics, the electrical stress is perpendicular to the interface, and alterations of interface shape combined with interfacial tension serve to balance the electric stress. Leaky dielectrics are different because free charge accumulated on the interface modifies the field. Viscous flow develops to provide stresses to balance the action of the tangential components of the field acting on interface charge.

9.2.1 BALANCE LAWS

The differential equations describing EHD arise from equations describing the conservation of mass and momentum, coupled with Maxwell's equations. To establish a context for the approximations inherent in the leaky dielectric model, it is necessary to analyze at a deeper level.[6]

Assumptions:

The hydrodynamic model consists of the Stokes equations without any electrical forces that are coupled to the electric field occurs at boundaries, hence forces from the bulk free charge must be negligible.

9.2.2 LEAKY DIELECTRIC MODEL

To summarize, the leaky dielectric EHD model consists of the following five equations. The derivation given here identifies the approximations in the leaky dielectric model. Except for the electrical body force terms, it is essentially the model proposed by Melcher and Taylor [7].

Assumptions:

(a) Dielectric fluid (poorly conductive liquid)

(b) Bulk charge = 0

(c) Axial motion

(d) Steady part

Fluid movement and a formation of a current balance equation must allow the following considerations [6,9]:

(a) Electromechanical coupling: fluid-to-fluid interface.

(b) Normal stress balances by changing the shape of the interface and interfacial tension.

(c) If electric field $\neq 0$, tangential stress balances by viscous force and causes the fluid motion.

(d) Liquid bulk is considered as quasi-neutral and free charges confined by a very thin layer under the liquid–gas interface.

Maxwell stress depends on electrostatic phenomena and hydrodynamic behavior. Hence, the electrical phenomena are described by [6–7]:

$$\nabla \varepsilon \varepsilon_0 E = \rho^e \tag{1}$$

and

$$\nabla \times E = 0 \tag{2}$$

where $\varepsilon \varepsilon_0$, ρ^e, and E are the ratio of the dielectric permeability, local free charge density, and electric field strength, respectively.

The relationship between Maxwell stresses and the electrical body force is to suppose that electrical forces exerted on free charge and charge dipoles are transferred directly to the fluid.[7,10]

$$P = NQd \tag{3}$$

where

P—polarization, Q—dipole charge, and d—orientation.

Polarization depends on the volumetric charge density (ρ^P) and surface charge density.

$$P = N\alpha\varepsilon\varepsilon_0 E \quad \text{and} \quad \nabla P = -\rho^P \tag{4}$$

The Coulomb force (due to free charge) $= \rho^F E$ $\qquad$ (5)

Total electrical force per unit volume

$$= \rho^F E + P.\nabla E = \nabla(\varepsilon\varepsilon_0 EE - \frac{1}{2}\varepsilon\varepsilon_0 E.E\delta) \tag{6}$$

where δ—two-dimensional vector, which balances by pressure gradient:

$$-\nabla P^* + \rho^F E + P.\nabla E = 0 \tag{7}$$

And the isotropic effect of E is

$$P^* = P + \frac{1}{2}\varepsilon_0 \left[\varepsilon - 1 - \rho \left(\frac{\partial \varepsilon}{\partial \rho} \right)_T \right] E.E \tag{8}$$

Therefore,

$$-\nabla P + \nabla.(\underbrace{\varepsilon\varepsilon_0 EE - \frac{1}{2}\varepsilon\varepsilon_0 \left[1 - \frac{\rho}{\varepsilon}\left(\frac{\partial \varepsilon}{\partial \rho} \right)_T \right] E.E\delta}_{\text{Maxwell stress tensor} = \sigma^M}) = 0 \tag{9}$$

The equation of motion of an incompressible Newtonian fluid of uniform viscosity [6] is

$$\rho \frac{\partial u}{\partial t} = -\nabla P + \nabla.\sigma^M + \mu \nabla^2 u \tag{10}$$

The electrical stresses emerge as body forces due to a nonhomogeneous dielectric permeability and free charge:

$$\rho \frac{\partial u}{\partial t} = -\nabla \left[p - \frac{1}{2}\varepsilon_0 \rho \left| \frac{\partial \varepsilon}{\partial \rho} \right|_T E.E \right] - \frac{1}{2}\varepsilon_0 E.E\nabla \varepsilon + \mu \nabla^2 u \tag{11}$$

EHD motions are driven by the electrical forces on boundaries or in the bulk. The net Maxwell stress at a sharp boundary has the normal and tangential components [7].

$$\left[\sigma^M.n \right].n = \frac{1}{2} \left\| \varepsilon\varepsilon_0 (E.n)^2 - \varepsilon\varepsilon_0 (E.t_1)^2 - \varepsilon\varepsilon_0 (E.t_2)^2 \right\|$$

$$\left[\sigma^M.n \right].t_i = qE.t_i \tag{12}$$

Finally, the equations are written in dimensionless variables using the scales defined. The equation of motion is for nonhomogeneous fluids with electrical body forces. The hydrodynamic boundary conditions, continuity of velocity and stress, including the viscous and Maxwell stress, are assumed [6].

$$\frac{\tau_\mu}{\tau_p}\frac{\partial u}{\partial t} + \mathrm{Re}\, u.\nabla u = -\nabla p - \frac{1}{2}E.E\nabla\varepsilon + \nabla.(\varepsilon E)E + \nabla^2 u \quad \& \quad \nabla.u = 0 \qquad (13)$$

$$\nabla.\sigma E = 0 \qquad (14)$$

$$\frac{\tau_c}{\tau_p}\frac{\partial q}{\partial t} + \frac{\tau_c}{\tau_F}\left[u.\nabla_s q - qn.(n.\nabla)u\right] = \left\|-\sigma E\right\|.n \qquad (15)$$

$$\left\|\varepsilon E\right\|.n = q \qquad (16)$$

$$\left[\sigma^M.n\right].n = \frac{1}{2}\left\|\varepsilon(E.n)^2 - \varepsilon(E.t_1)^2 - \varepsilon(E.t_2)^2\right\|$$
$$\left[\sigma^M.n\right].t_i = qE.t_i \qquad (17)$$

9.3   THERMO-ELECTRO-HYDRODYNAMIC MODEL

EHD model is a simple model without considering the thermal effect. In 2004, researchers [8] considered the coupling effects of thermal, electrical, and hydrodynamics. A complete set of balance laws governing the general thermo-electro-hydrodynamics flows has been derived before. It consists of modified Maxwell's equations governing electrical field in a moving fluid, the modified Navier–Stokes equations governing heat and fluid flow under the influence of electric field, and constitutive equations describing behavior of the fluid (Figure 2).

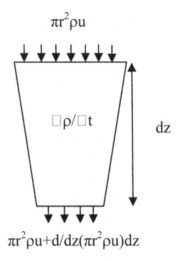

$\pi r^2 \rho u$

$\Box \rho / \Box t$

dz

$\pi r^2 \rho u + d/dz(\pi r^2 \rho u)dz$

FIGURE 2   Schematic of a control volume.

Governing equations without thermal effects are [11,12]:

(1) Mass balance: $\nabla.u = 0$ (18)

(2) Linear momentum balance: $\rho(u.\nabla)u = \underbrace{\nabla T^m + \nabla T^e}_{viscous\ and\ electric\ forces}$ (19)

(3) Electric charges balance: $\nabla.J = 0$ (20)
Modified Navier–Stokes equations are [8]:

$$\frac{\partial q_e}{\partial t} + \nabla.J = 0 \tag{21}$$

$$\rho \frac{Du}{Dt} = \nabla.t + \rho f + q_e E + (\nabla E).P \tag{22}$$

$$\rho c_P \frac{Du}{Dt} = Q_h + \nabla.q + J.E + E.\frac{DP}{Dt} \tag{23}$$

This set of conservation laws constitute a closed system when it is supplemented by appropriate constitutive equations for the field variables such as polarization. The most general theory of constitutive equations determining the polarization, electric conduction current, heat flux, and Cauchy stress tensor. [8,13]

$$P = \varepsilon_p E \tag{24}$$

$$J = kE + \sigma u + \sigma_T \nabla T \tag{25}$$

$$q = K \nabla T + K_E E \tag{26}$$

$$t = -\tilde{P}\underline{\underline{I}} + \eta\left[\nabla\underline{\upsilon} + (\nabla\underline{\upsilon})'\right] \tag{27}$$

Here, coefficients ε_p (electric susceptibility), μ_m, k, σ, σ_T, K, K_E, and η are material properties and depend only on temperature in the case of an incompressible fluid.

9.4 MATHEMATICAL MODEL

An unsteady flow of an infinite viscous jet pulled from a capillary orifice and accelerated by a constant external electric field (Table 1).[8]

(1) Mass conservation: $\dfrac{\partial}{\partial t}(r^2) + \dfrac{\partial}{\partial z}(r^2 u) = 0$ $\hfill (28)$

(2) Charge conservation:

$$\frac{\partial}{\partial t}\left(2\pi r(\sigma + \varepsilon_p E)\right) + \frac{\partial}{\partial z}(2\pi r(\sigma + \varepsilon_p E)u + \pi r^2 kE + \pi r^2 \sigma_T \frac{\partial T}{\partial z}) = 0 \tag{29}$$

where σ is the surface charge density and E the electric field in the axial direction.

TABLE 1   Different types of current

Formula	Current
$J_c = \pi r^2 kE$	The ohmic bulk conduction current
$J_s = 2\pi r \sigma u$	Surface convection current
$J_T = \pi r^2 \sigma_T \dfrac{\partial T}{\partial z}$	Current caused by temperature gradients

(1) The Navier–Stokes equations becomes [8,11,13]

3.1. Momentum equation:

$$\frac{\partial u}{\partial t} + u\frac{\partial u}{\partial z} = -\frac{1}{\rho}\frac{\partial p}{\partial z} + g + \frac{2\sigma E}{\rho r} + \frac{1}{r^2}\frac{\partial \tau}{\partial z} + \frac{1}{r^2}\varepsilon_p E\frac{\partial E}{\partial z} \tag{30}$$

3.2. Energy equation:

$$\rho c_p\left(\frac{\partial T}{\partial t} + u\frac{\partial T}{\partial z}\right) = Q + \frac{\partial}{\partial z}(k\frac{\partial T}{\partial z} + k_E E + 2\pi r\sigma u + \pi r^2 kE + \pi r^2 \sigma_T \frac{\partial T}{\partial z})E + \varepsilon_p E\left(\frac{\partial E}{\partial t} + u\frac{\partial E}{\partial z}\right) \tag{31}$$

Internal pressure of the fluid:

$$p = K\gamma - \frac{\varepsilon - \varepsilon_0}{g\pi}E^2 - \frac{2\pi}{\varepsilon_0}\sigma^2 \tag{32}$$

Twice the mean curvature of the interface:

$$K = \frac{1}{R_1} + \frac{1}{R_2} \tag{33}$$

where R_1 and R_2 are the principal radii of curvature.

9.5 CONSTITUTIVE EQUATIONS

Rheologic behavior of many polymer fluids can be described by power-law constitutive equation in the form [14]:

$$\tau = \mu_0\frac{\partial u}{\partial z} + \sum_{n=1}^{m} a_n\left(\frac{\partial u}{\partial z}\right)^{2n+1} \tag{34}$$

In dielectrics the charges are not completely free to move, but the positive and negative charges that compose the body may be displaced in relation to one another when a field is applied so the following equations can be used for solving governing Eq. (8).

$$q_\rho = -\nabla.P \quad \text{and} \quad P = \varepsilon_p E \quad E \tag{35}$$

9.6 CONCLUSION

A complete thermo-electro-hydrodynamic model considers the coupling effects of thermal and electric field. The disadvantage of this model is that it is too complex for numerical analysis. Therefore, a one-dimensional thermo-electro-hydrodynamic model, which can be applied to numerical study, is derived. The model can offer in-depth insight into physical understanding of many complex phenomena, which cannot be fully explained experimentally. It is a powerful tool for controlling other physical characters.

KEYWORDS

- **Electrospinning**
- **Electrohydrodynamics**
- **Thermo-electro-hydrodynamics**

REFERENCES

1. Greiner, A.; and Wendorff, J. H.; Electrospinning: a fascinating method for the preparation of ultrathin fibers. *Angew. Chemie. Int. Ed.* **2007,** *46(30),* 5670–5703.
2. Subbiah, T.; et al. Electrospinning of nanofibers. *J. Appl. Polym. Sci.* **2005,** *96(2),* 557–569.
3. Ohgo, K.; et al. Preparation of non-woven nanofibers of bombyx mori silk, samia cynthia ricini silk and recombinant hybrid silk with electrospinning method. *Polym.* **2003,** *44(3),* 841–846.
4. Lukáš, D.; et al. Physical principles of electrospinning (electrospinning as a nano-scale technology of the twenty-first century). *Text. Prog.* **2009,** *41(2),* 59–140.

5. Huang, Z. M.; et al. A review on polymer nanofibers by electrospinning and their applications in nanocomposites. *Compos. Sci. Technol.* **2003**, *63(15)*, 2223–2253.

6. Saville, D. A.; Electrohydrodynamics: the Taylor–Melcher leaky dielectric model. *Ann. Rev. Fluid Mech.* **1997**, *29(1)*, 27–64.

7. Melcher, J. R.; and Taylor, G. I.; Electrohydrodynamics: a review of the role of interfacial shear stresses. *Ann. Rev. Fluid Mech.* **1969**, *1(1)*, 111–146.

8. Wan, Y. Q.; Guo, Q.; and Pan, N.; Thermo-electro-hydrodynamic model for electrospinning process. *Int. J. Nonlinear Sci. Numer. Simul.* **2004**, *5(1)*, 5–8.

9. Zhang, J.; and Kwok, D. Y.; A 2-D lattice Boltzmann study on electrohydrodynamic drop deformation with the leaky dielectric theory. *J. Comput. Phys.* **2005**, *206(1)*, 150–161.

10. Griffiths, D. J.; and College, R.; Introduction to Electrodynamics. Upper Saddle River, NJ: Prentice Hall; **1999**, *3*, 576 p.

11. He, J. H.; and Liu, H. M.; Variational approach to nonlinear problems and a review on mathematical model of electrospinning. *Nonlinear Anal.: Theory, Methods, Appl.* **2005**, *63(5)*, e919–e929.

12. Spivak, A. F.; and Dzenis, Y. A.; Asymptotic decay of radius of a weakly conductive viscous jet in an external electric field. *Appl. Phys. Lett.* **1998**, *73(21)*, 3067–3069.

13. Xu, L.; Wang, L.; and Faraz, N.; A thermo-electro-hydrodynamic model for vibration-electrospinning process. *Therm. Sci.* **2011**, *15(suppl. 1)*, 131–135.

14. Munir, M. M.; et al. Scaling law on particle-to-fiber formation during electrospinning. *Polym.* **2009**, *50(20)*, 4935–4943.

CHAPTER 10

LIQUID JET–AIR INTERFACE IN ELECTROSPINNING PROCESSES

A. K. HAGHI and A. POURHASHEMI

CONTENTS

10.1   INTRODUCTION

Nanotechnology is considered one of the most promising technologies for the twenty-first century. Nanotechnology is used when either nanoscaled materials are produced (defined by their thickness, particle size, or other structural features) or the nature of a process involves the use of nanoscaled materials. Nanotechnology is considered the future in manufacturing technology that will result in products that are lighter, stronger, cleaner, less expensive, and more precise. Research and development in nanotechnology are directed toward understanding and creating improved materials, devices, and systems that exploit these new properties.[1]

Electrospinning, discovered in the early 1900s, as a branch of nanotechnology, has attracted a lot of interest as a novel technique that is a very simple and inexpensive way of manufacturing continuous nanofibers ranging from less than 10 nm to over 1 μm in diameter. The ability to electrospin fibers from diverse classes of material has resulted in a huge range of potential applications and growing interest in the process by researchers worldwide. It has potential application in filters, tissue engineering scaffolds, wound dressings, drug delivery materials, biomimetic materials, composite reinforcement, protective clothing, electronics, implants, agriculture, and many other areas.[2]

The electrospinning process (Figure 2) is based on a simple concept of creating nanofibers through an electrically charged jet of polymer solution or polymer melt, as shown in Figure 1.

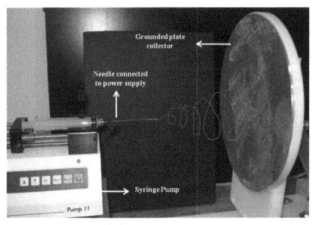

FIGURE 1   Standard electrospinning setup.

FIGURE 2 The steps of electrospinning process.

The principle behind electrospinning is relatively simple. When voltage is initially applied to the fluid, the droplet at the nozzle gets distorted into the shape of a cone. The final conical shape is known as the Taylor cone. These changes are due to the conflict between the increasing solution charge and its surface tension. When the applied voltage is sufficient, the electrostatic force in the polymer and solvent molecules have enough charge to overcome surface tension and a stream is ejected from the tip of the Taylor cone. The solution is drawn as a jet toward an oppositely charged collecting plate, which causes the charged solution to accelerate toward the collector. The solvent gradually evaporates, and a charged, solid polymer fiber accumulates on the collection plate [3,4] (Figure 2).

Electrospinning (Figure 1) seems to be a relatively simple process for producing nanofibers as it utilizes a few readily available components. On closer examination, it is however clearly evident that successful electrospinning involves an understanding of the complex interaction of electrostatic fields, properties of polymer solutions, and component design and system geometry. The application of a high voltage to the solution has two broad regions of impact. The first is "internal" with the Coulombic forces between ions of the same charges causing the ions to disperse. Countering this is the surface tension force, which tends to minimize the liquid surface area and surface energy based on the cohesiveness of the liquid molecules, causing them to get collected together. The former force causes the fluid to disassociate, whereas the latter tends to consolidate the fluid stream. The latter force is "external" and is the field generated between the needle and the collector, which causes the charged solution to travel from needle toward the grounded collector.[5]

During this process, the jet might show some unstable behavior. Reneker et al. [6] described that the bending segments of a loop in the instability region of the jet can suddenly develop a new bending instability, similar to, but at a smaller scale than, the first. However, the most common type of instability is a spiral formation with an increasing diameter. This is called the "whipping mode."

Studying the dynamical behavior of the jet is of interest for the development of a possible control system. Such a study should focus on the various instabilities that are possible in a jet. Modeling and simulations will give a better understanding of electrospinning jet mechanics. In addition, the effect of secondary external field can be studied using simulation studies. Till date, two major modeling zones have been identified. These zones are as follows: (a) the zone close to the capillary outlet where an axisymmetric jet exits and thins down and (b) the whipping instability zone where the jet spirals and accelerates toward the collector plate.[7]

As discussed, modeling and simulation of electrospinning jet will give a better understanding of the process. Hence, in this project, the electrospinning jet and the whipping instability will be modeled and simulated.

10.2   MATHEMATICAL MODEL

To understand the mathematical model, few factors must be considered about the modeling of the system. The model of the fluid jet is described as a system of beads that are interconnected by viscoelastic elements. These elements are also called filaments. The forces discussed below act on each bead in the system.

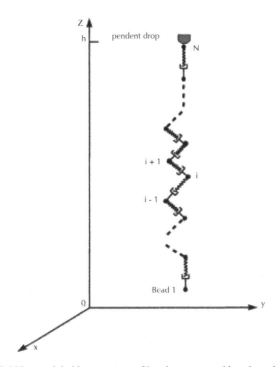

FIGURE 3   Fluid jet modeled by a system of beads, connected by viscoelastic elements.

In Figure 3, one bead, i, of the system is shown. The forces acting on bead i are dependent on the position of the other beads that are present in the system, as bead i is interconnected with other beads. If the forces acting on bead i are known, then the acceleration of this bead can be calculated by the equations of motion, using Newton's second law.

Maxwell constitutive equation was first applied by Reneker et al. [6] in 2000. Consider an electrified liquid jet in an electric field parallel to its axis. They modeled a segment of the jet after a viscoelastic dumbbell. They used a Gaussian electrostatic system of units. According to this model, each particle in the electric field exerts a repulsive force on another particle.

The following were three main assumptions [6,8]:

(1) The background electric field created by the generator is considered static.

(2) The fiber is a perfect insulator.

(3) The polymer solution is a viscoelastic medium with constant elastic modulus, viscosity, and surface tension.

10.2.1   VISCOELASTIC FORCES

The viscoelastic force acting on bead i is due to a stress s. This stress, which acts on an element between two beads, models a viscoelastic Maxwell liquid jet. The Maxwell model describes a spring-damper system and the equation of motion of this system is given by [6]

$$\frac{d\varepsilon}{dt} = \frac{1}{G}\frac{d\sigma}{dt} + \frac{\sigma}{\mu} \tag{1}$$

where ε is the strain of the element, G and μ are the elastic modulus and the viscosity of the material, respectively, and σ is the stress on the element.

To adapt (1) to the mathematical model, again consider bead i. This bead is connected by viscoelastic elements to two other beads, bead $i + 1$, which is above bead i, and bead $i - 1$, which is below. This model is as shown in Figure 3. The stresses between bead i and the two beads can then be defined by the two differential equations as follows [9]:

$$\frac{d\sigma_{ui}}{dt} = G\frac{1}{l_{ui}}\frac{dl_{ui}}{dt} - \frac{G}{\mu}\sigma_{ui} \tag{2}$$

$$\frac{d\sigma_{bi}}{dt} = G\frac{1}{l_{bi}}\frac{dl_{bi}}{dt} - \frac{G}{\mu}\sigma_{bi} \tag{3}$$

The distances between the beads and thus the length of the filaments are given by l_{ui} and l_{bi}, where the subscript ui denotes the filament between bead i and $i + 1$, while the subscript bi denotes the filament between bead i and $i - 1$. These distances are given by the Pythagorean theorem, as described in the following two equations:

$$l_{ui} = \sqrt{(x_{i+1} - x_i)^2 + (y_{i+1} - y_i)^2 + (z_{i+1} - z_i)^2} \tag{4}$$

$$l_{bi} = \sqrt{(x_i - x_{i-1})^2 + (y_i - y_{i-1})^2 + (z_i - z_{i-1})^2} \tag{5}$$

10.2.2 COULOMBIC FORCES

In bead i, another factor that has to considered is the Coulomb force. This force is influenced by the position of each bead in the system. Other forces on any bead i are caused by the location of the two other beads $i - 1$ and $i + 1$. However, the total Coulomb force (Figure 4) on bead i is a summation of the Coulomb forces acting on i caused by every bead present in the system, except for bead i itself.

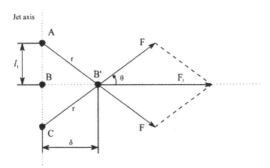

FIGURE 4 Earnshaw instability, caused by the Coulomb forces, leading to the bending of an electrified jet.

The magnitude of the Coulomb force acting on charge q_1 due to the presence of a charge q_2 is then given by the following equation:

$$F = \frac{q_1 q_2}{R^2} \tag{6}$$

Bead B is placed between the two beads A and C. The charges of beads A and C create a Coulomb force having magnitude $F = \frac{e^2}{r^2}$. This force then pushes from both sides against bead B.

$$F_1 = 2F\cos(\theta) = 2\frac{e^2}{r^2}\cos(\theta) = 2\frac{e^2}{r^3}\delta \tag{7}$$

The total Coulomb force acting on bead i can now be derived by calculating the sum of the Coulomb forces generated by all the beads present in the system [10]:

(a) Assume that the charges of all beads in the system are equal.

(b) The distance R in (6) can be defined as the distance $R_{i,j}$ between bead i and a bead j. This distance can be derived from the Pythagorean theorem:

$$R_{i,j} = \sqrt{(x_i - x_j)^2 + (y_i - y_j)^2 + (z_i - z_j)^2}$$

(8)

With $(i, j = 1,\ldots, N, i \neq j)$.

$$F_c = \sum_{\substack{j=1 \\ j \neq i}}^{N} \frac{e^2}{R_{ij}^3} \left[(x_i - x_j)\hat{i} + (y_i - y_j)\hat{j} + (z_i - z_j)\hat{k} \right]$$

(9)

10.2.3   SURFACE TENSION FORCE

The surface tension force tends to restore the jet into its rectilinear shape. The surface tension force is derived from the surface tension coefficient. This coefficient is typically defined as a force along a line of unit length. If this is multiplied by the surface of the jet and its curvature (Figure 5), then the surface tension force for a segment can be calculated. The surface tension force acting on the ith bead is then given by [9]

$$F_{st} = -\frac{\alpha \pi a_{av}^2 k_i}{\sqrt{x_i^2 + y_i^2}} \left[ix_i + jy_i \right]$$

(10)

where α is the surface tension coefficient and a is the average radius of a_{av} segment of the jet between two beads, given by

$$a_{av}^2 = \frac{(a_{ui} + a_{bi})^2}{4}$$

(11)

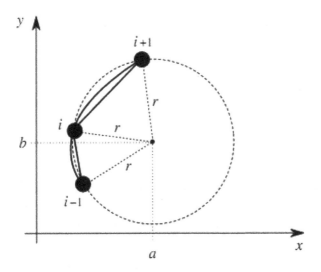

FIGURE 5   Estimation of the curvature.

10.2.4   ELECTRIC FORCE

The electric force acting on bead i is derived from the potential difference between the pendent drop and the collector. This difference yields an electric field. As the potential difference in this system is only applied in the z-direction, this electric field is a 1D field. The electric field can also be defined as the force the bead experiences per unit charge. Thus, the force can then be given by [9,10]

$$F_{e} = -e\frac{V_0}{h}\hat{k} \tag{12}$$

where h is the distance between the pendent drop and the collector.

10.2.5   EQUATION OF MOTION

As all the different forces acting on each bead are now defined, Newton's second law can be used to derive the equations of motion [9]:

$$m\frac{d^2 r_i}{dt^2} = \sum_{\substack{j=1 \\ j \neq i}}^{N} \frac{e^2}{R_{ij}^2}(r_i - r_j) - e\frac{V_0}{h}\hat{k} + \frac{\pi a_{ui}^2(\bar{\sigma}_{ui} + G\ln(l_{ui}))}{l_{ui}}(r_{i+1} - r_i) - \frac{\pi a_{bi}^2(\bar{\sigma}_{bi} + G\ln(l_{bi}))}{l_{bi}}(r_i - r_{i-1})$$

$$-\frac{\alpha\pi a_{av}^2 k_i}{\sqrt{(x_i^2 + y_i^2)}}[ix_i + jy_i] \tag{13}$$

with $i = 1,\ldots, N$. This equation represents the mathematical model in 3D space.

10.3 MATLAB SCRIPT

As all the differential equations are defined, the MATLAB script can be designed. There are several ways to do this. In each case, the main feature in the script is its ability to cope with the introduction of new beads into the system and the removal of beads that have reached the collector plate.

The disadvantages of this program are as follows:

- New beads are introduced into the system and removed if they reach the collector: This is the most complex situation, as the state vector has a varying size. Moreover, another difficulty is that beads that have reached the collector must be removed from the state vector.
- New beads are introduced into the system and are reintroduced into the system after reaching the collector: In this manner, the amount of beads and thus the amount of differential equations stay bounded within a specified amount of beads. The integrations start with one bead and the system grows after the introduction of new beads each time. When the first bead reaches the collector, it will be reintroduced into the system, which leaves the size of the state vector unchanged.
- Integration starts with specified amount of beads that are reintroduced into the system after reaching the collector: This is the simplest way to describe the system, as the size of the state vector is specified before the integration starts and does not change during the simulation.

10.4 CONCLUSION

The investigations describe how to create a model that can simulate the electrospinning process and how to study the instability of the fluid jet. The different versions of the MATLAB script are used for these simulations. The electrospinning process is dependent on a lot of different parameters. Changing these parameters will lead to changes in the process. The influence of the viscosity, the surface tension, the elastic modulus, the initial jet radius, and the applied voltage can also be studied.

KEYWORDS

- **Liquid jet–air interface**
- **Nanofiber**
- **Electrospinning**

REFERENCES

1. Sawhney, A. P. S.; et al. Modern applications of nanotechnology in textiles. *Text. Res. J.* **2008,** *78(8),* 731–739.
2. Brown, P. J.; and Stevens, K.; Nanofibers and Nanotechnology in Textiles. Florida: CRC Press Boca Raton; **2007,** 544 p.
3. Pham, Q. P.; Sharma, U.; and Mikos, A. G.; Electrospinning of polymeric nanofibers for tissue engineering applications: a review. *Tissue Eng.* **2006,** *12(5),* 1197–1211.
4. Huang, Z. M.; et al. A review on polymer nanofibers by electrospinning and their applications in nanocomposites. *Compos. Sci. Technol.* **2003,** *63(15),* 2223–2253.
5. Lukáš, D.; et al. Physical principles of electrospinning (electrospinning as a nano-scale technology of the twenty-first century). *Text. Prog.* **2009,** *41(2),* 59–140.
6. Reneker, D. H.; et al. Bending instability of electrically charged liquid jets of polymer solutions in electrospinning. *J. Appl. Phys.* **2000,** *87,* 4531–4547.
7. Feng, J. J.; Stretching of a straight electrically charged viscoelastic jet. *J. Non-Newtonian Fluid Mech.* **2003,** *116(1),* 55–70.
8. Reneker, D. H.; et al. Electrospinning of nanofibers from polymer solutions and melts. *Adv. Appl. Mech.* **2007,** *41,* 43–195.
9. Van Vught, R.; Simulating the Dynamical Behaviour of Electrospinning Processes, in Department of Mechanical Engineering. Eindhoven: Eindhoven University of Technology; **2010,** 65 p.
10. Darsi, T.; Mathematical models of bead-spring jets during electrospinning for fabrication of nanofibers. *Walailak J. Sci. Technol. (WJST).* **2012,** *9(4),* 287–296.

INDEX

Printed in the United States
by Baker & Taylor Publisher Services